Telecommunications in Asia

Policy, Planning and Development

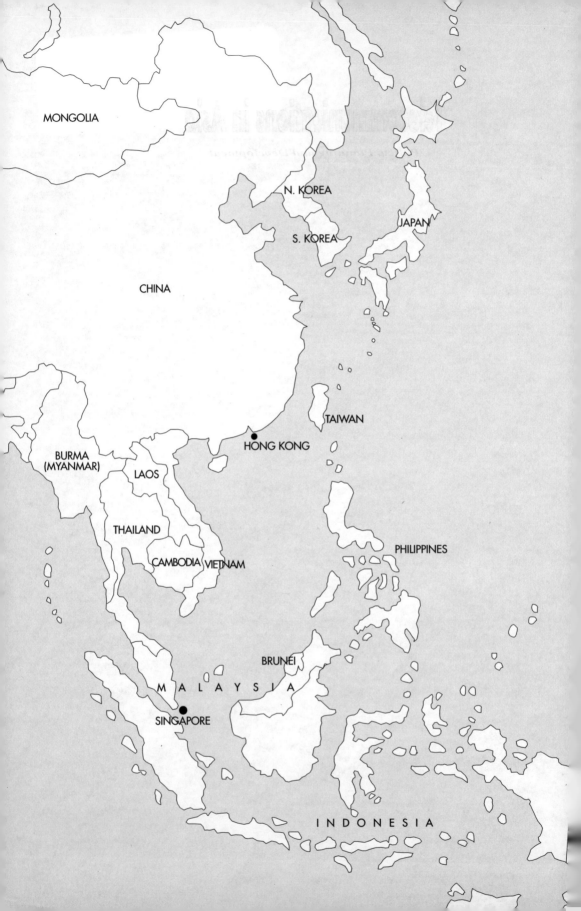

Telecommunications in Asia

Policy, Planning and Development

Edited by John Ure

Contributors

Andrew Harrington

Nick Ingelbrecht

Peter Lovelock

Susan Schoenfeld

John Ure

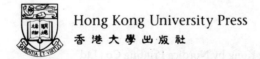

Hong Kong University Press
香港大學出版社

Hong Kong University Press
139 Pokfulam Road, Hong Kong

© Hong Kong University Press 1995

ISBN 962 209 383 3

All rights reserved. No portion of this
publication may be reproduced or
transmitted in any form or by any means,
electronic or mechanical, including
photocopy, recording, or any information
storage or retrieval system, without
permission in writing from the publisher.

Printed in Hong Kong by Nordica Printing Co., Ltd.

Contents

- Preface and Acknowledgements   vii

- Contributors   ix

- **Chapter 1**   1
  Introduction

- **Chapter 2**   11
  Telecommunications in China and the Four Dragons
  John Ure

- **Chapter 3**   49
  Telecommunications in ASEAN and Indochina
  John Ure

- **Chapter 4**   81
  Companies and Capital in Asia-Pacific Telecommunications
  Andrew Harrington

- **Chapter 5**   111
  Information Highways and the Trade in
  Telecommunications Services
  Peter Lovelock

- **Chapter 6**   147
  The Broadcast Media Markets in Asia
  Peter Lovelock and Susan Schoenfeld

- **Chapter 7** 193
  Asia — The Supplier's Dilemma
  Nick Ingelbrecht

- **Chapter 8** 235
  Conclusion

- Appendix 243

- Glossary 249

- References 263

- Index 277

Preface and Acknowledgements

This book grew out of research which, for the editor, began in 1989. In that year I arrived in Hong Kong on sabbatical leave from the Department of Applied Economics, Polytechnic of East London (now the University of East London) with every intention of studying telecommunications in Hong Kong for one year. The assumption was that this industry, a paradigm of technological change, could be understood in its essentials and written up as a case study after twelve months of field work — see Ure (1989, 1992). What in fact I learned was that the changes overtaking the industry were quintessentially a microcosm of the micro-electronics revolution, with profound implications for the internal management of capital, labour and technology — see Ure, 1994 — and equally profound implications for the regulation of the industry as all the boundary assumptions as to where the telecommunications industry ended and where the computing and media industries began dissolved into history.

It became equally evident that, as a result, the role of telecommunications in national, economic and social development of the Southeast and East Asian region was changing and growing in significance. Governments throughout the region were shifting their policies, seeking ways to encourage investment in modern networks, but simultaneously ambivalent about the potential for wider access to the means of communication, especially where that allowed citizens access to sources of news and information not controlled by national regulations.

Tracking these contradictory developments becomes a way into understanding the dynamics, or political economy, of policy making in the region. There are many aspects involved: from the planning and development of the telecommunications networks themselves, to methods of financing and policies towards market reforms, to regulatory issues and the challenge of transborder information flows and multi-media developments that defy traditional industry boundaries, to the basic issues of equipment procurement, standards issues and the need for technology transfer to promote the development of a local manufacturing capability. My research had focused very much on the policy and regulatory aspects of network development, so it made sense to turn to research specialists for expertise in the other areas. The result, after eighteen months, is this book.

Each one of the authors has been helped and assisted by many generous friends and acquaintances throughout the region, just too many to name individually. They include government officials and regulators, management and staff of telecommunications companies, journalists and media people, consultants and market analysts, scholars and researchers from many countries. None is responsible for any of the views expressed in this book, or for any errors that may have slipped in — for which I, as editor, must alone be held to account — but we hope this book is a tribute to them.

But the following acknowledgments are in order. First, since 1993 I have been the Director of the Telecommunications Research Project at the Centre of Asian Studies, University of Hong Kong, which is funded by a generous grant from the HongKong Telecom Foundation. The research is all in the public domain, with no strings attached to the grant, but without it the research and the time for writing and editing this book would not have come easy. I can only hope such an enlightened funding policy may serve as a model for other parts of the region, where universities and research institutes are desperately short of funds. I am especially indebted to Professor Edward Chen Kwan-yiu, Director of the Centre of Asian Studies, to Ms Coonoor Kripalani-Thadani, Assistant to the Director, and to the staff of the Centre for their generous support for the Project. Second, I wish to thank Ms Ester Castaneda for the many painstaking and good-humoured hours she put into classifying my research material and typing and checking manuscripts, and to Ms Joyce Nip, of the Department of Journalism, Baptist University of Hong Kong, for her critical eye and editing advice. Finally, to Dr Ben Petrazzini of the Hong Kong University of Science and Technology for his assiduous and astute comments on an early draft of the chapters, and to the staff of Hong Kong University Press for their close collaboration, advice and support.

John Ure
Hong Kong, 11 March 1995

Contributors

John Ure is an economist specializing in the study of telecommunications and related information technologies. He obtained his B.Sc. (Econ) from the University of Hull, M.Sc. (Econ) from Birkbeck College, University of London, and Ph.D. (Econ) from the Polytechnic of East London. He is the Director of the Telecommunications Research Project at the Centre of Asian Studies, the University of Hong Kong, where he supervises a number of postgraduate students. His related publications include 'Telecommunications, With Chinese Characteristics' in the China Special Issue of *Telecommunications Policy* April 1994; 'Telecommunications: Hong Kong and China After 1997' in J.Cheng and S.Lo eds. *From Colony to SAR: Hong Kong's Challenge Ahead* (The Chinese University Press: Hong Kong, 1995); and 'New Technology and Hong Kong Telephone Workers' in *New Technology, Work and Employment*, volume 8.2, September 1994. He is a member of the Hong Kong Consumer Council Steering Group on Telecommunications Policy, and in 1994 was a member of the Hong Kong delegation to the APEC Telecommunications Working Group.

Andrew Harrington is a Director in Research at Salomon Brothers Hong Kong Limited. Dr Harrington follows the telecommunications industry in Asia Pacific region on behalf of institutional investors worldwide, and was recently ranked by Institutional Investor as the only 'Sector Star' in telecommunications in Asia. He has also been ranked first in telecommunications research in the Asiamoney survey of institutional investors. He had previously followed the European telecommunications industry for Salomon Brothers in London. Dr Harrington has been based in Asia since 1991 and has written extensively on the industry. He is a well known speaker at industry conferences and is widely quoted in the international press. Dr Harrington received an M.A. from Cambridge University in Natural Sciences and a D.Phil. in Solid State Physics from Oxford University.

Nick Ingelbrecht is a journalist specializing in telecommunications, cable and satellite broadcasting. He has worked in and reported on the telecommunications industry for more than 15 years, latterly focusing on the Asia Pacific markets from his base in Hong Kong, where he is a regular contributor to several leading industry journals, including *Communications Week International* and *Asia Pacific Telecommunications*. He is an Associate of the Telecommunications Research Project at the University of Hong Kong's Centre of Asian Studies and holds a B.A. from the University of Hull in England.

Peter Lovelock is a Ph.D. candidate in political science at the University of Hong Kong. He obtained his undergraduate degree (in economics and political science) from Sydney University in Australia. He has written extensively on both telecommunications and media industries in Asia and acted as an adviser to the Australian government in the establishment of their multimedia and pay-TV initiatives. He is an Associate the Telecommunications Research Project at the University of Hong Kong's Centre of Asian Studies and a participant at APEC Telecommunications Working Group. His Ph.D. dissertation is an analysis of the trade and telecommunications relationships in Northeast Asia.

Susan J. Schoenfeld is the President of Advisors for International Media (AIM) Asia Ltd, a media and telecommunications consultancy based in Hong Kong. She is an Associate of the Telecommunications Research Project at the University of Hong Kong's Centre of Asian Studies, where she is a Visiting Scholar. She came to Hong Kong, on a Fulbright Fellowship, after practising media and telecommunications law in New York City for five years. She has taught media and telecommunications law and policy at the Chinese University of Hong Kong. She is the Senior Advisor to the Media Roundtable of PECC and has served as a US representative in Hong Kong to the PECC and APEC telecommunications working group activities. Ms Schoenfeld has written for publications in the US, Europe and Asia. She graduated with honors from Wellesley College and received her Juris Doctor from Columbia University Law School.

Introduction

Despite the explosive growth of telecommunications taking place throughout Asia — see the annex to this chapter — to date it has received surprisingly little detailed analysis. This is all the more surprising as the planning and policy developments are not recent. Since the early 1980s the states of Southeast and East Asia have been consciously planning for the shift towards an information technology-based economy, and as Kaplinsky (1987, p.47) notes, during the 1980s the growth of telecommunications in the developing countries outstripped that in the developed countries, with the whole of Asia leading the way with an annual average increase in expenditure of US$10 billion in 1982–87. What is phenomenal in the 1990s is the pace of acceleration. China's plan for posts and telecommunications commits 80 billion yuan, the equivalent of almost US$10 billion for 1995 alone.

This book aims to provide such an analysis, with a focus on practical issues, such as policy and regulation, finance, trade-in-telecommunications and related information and media services, and equipment supply and equipment standards. The countries covered by the chapters which follow are the developing economies of Southeast and East Asia, from Indochina and the ASEAN states, to China and the dragon economies of Hong Kong, Singapore, South Korea and Taiwan.[1]

Two underlying propositions run through this book. The first acknowledges the common factor of the world market as a driver of telecommunications across the region, but argues that local conditions within each nation-state mediate the policy-making process, and attention must be given to this local context if the dynamics of Asian telecommunications is to be understood. The second highlights the role of local Asian capital, arguing that it is wrong to see Asian economies simply as large markets for Western and Japanese investment and equipment sales, and wrong to see Asian states as simply reacting against the invasion of Western and Japanese cultural and information products, such as satellite TV. Rather, Asian states are encouraging a vigorous outward expansion of their own industries, meeting the Western and Japanese multinationals at the regional level. The result is a volatile mixture of alliances and rivalries, cooperation and competition among local Asian companies expanding into one another's territories and forming equally volatile relationships with the multinationals.

Thirty per cent of the world's 5.5 billion population live in the economies covered by this book — in fact 25 per cent live in China and Indonesia — yet, as the *World Telecommunications Development Report* (ITU, 1994) indicates, they have only 8 per cent of the world's main telephone lines. If the four dragon economies are excluded — around 70 million people — the figure falls to 3 per cent.[2] A decade ago the Maitland Commission's Report *The Missing Link* (ITU, 1984) revealed that just nine countries accounted for 75 per cent of the world's 600 million telephones. A decade later (ITU, 1994), fifteen countries accounted for 75 per cent of the world's 570 million telephone mainlines.[3] Clearly the Maitland Report's primary objective that 'by the early part of the next century virtually the whole of mankind should be brought within the easy reach of a telephone'[4] remains a good way off. Furthermore, this calculation does not take into account the fact that the richer industrial nations are also enjoying Maitland's secondary objective, 'the other services telecommunications can provide'.

What characterizes telecommunications development since Maitland is not just the accelerated growth of the basic network, but the unevenness of this growth, between and within economies. Some economies, like those of Hong Kong and Singapore, have pulled further ahead, while others, such as Indochina and the Philippines have fallen further behind. Within countries, global trading centres, such as the coastal cities of China — Guangzhou, Ziamen, Shanghai, Beijing and Tianjin — and the major commercial cities of Malaysia — Kuala Lumpur, Penang and Johor Bahru — and several of the region's capital cities — Bangkok, Jakarta and Manila — are fast modernizing their networks, transforming themselves (their central and business districts) into world cities. Meanwhile, the rural areas and smaller townships are frequently without access to the basic service.[5]

Uneven growth would be less significant were it not for the changing role of telecommunications and information technologies in modern economies. In the process of nation-state building going on in Asia, governments place high priority on modernization, which they closely identify with developing a national capability in applied science and new technologies. Telecommunications feeds this development in two ways. First, a local telecommunications equipment and components industry helps reduce the need for imports, and holds out the potential to grow into an export sector. To develop a capability in this area Asian economies require technology transfer of research and development in design and manufacturing, an issue taken very seriously by countries such as China and Malaysia. They also need a home market to supply, that is an expanding telecommunications network. Second, network infrastructure is vital for the information economy. No Asian country today wants to proceed slowly in developing its own version of the information superhighway.

Asia is not the United States, nor is it Europe; and the evolution of information technologies in Asia will take on its own characteristics. For example, given the difficulties of establishing common standards for computer programming of some Asian written languages, online modes of communication may be slow to diffuse, whereas other modes, such as 'scribble and send' using facsimile and digitized notebooks, short-messaging services, available by pagers and digital cellular phone handsets, and voice-messaging services by telephone, are already very popular. However, the technological trends and economic consequences of information technologies evident in the West and Japan will have similar implications for Asia, and in that sense the United States and Europe offer a mirror to the future.

What is that future? Two remarkably lucid and readable accounts of it, which map the changes overtaking the telecommunications industry, are Huber (1987) and Huber, Kellogg and Thorne (1993). And two of the important changes they flag are immediately relevant for Asia.

The first change overtaking the industry is the falling cost of switching and of intelligence in the network. As the cost of processors and memory chips declines, it becomes economical to build intelligent functions, including switching, into small remote units. They migrate along the transmission lines towards nodes, where networks interconnect, such as the PABX (private automation branch exchange, or company switchboard) and to user terminals, such as tone-dial handsets, facsimile machines and computers. This sets up the technological possibility of companies and other user groups developing their own virtual networks, of by-passing the public switched telephone network (PSTN) and the public switched data network (PSDN) — and their tariffs — by re-routing traffic. The traditional natural monopoly public network dissolves into a network of interconnections with private and new entry public networks (assuming, of course, regulation permits). In Asia this development is in its very early stages, and where it has occurred the necessary regulatory changes are still being worked out.

The second change is actually a revival of Marconi's[6] original vision of radio telecommunications serving people who are mobile — Marconi was thinking of ship-to-shore communications — rather than providing fixed point-to-point communications. Technologies like cellular mobile radio, pagers, CT2 (cordless telephone second generation), trunked radio, and now fixed cellular radio in the local loop (that is, the local telephone network) are already widespread in parts of Asia where fixed telephone lines are few on the ground. And the next generation of radio communications, which assigns numbers not to locations, but to people, are on the verge of introduction, alongside satellite mobile telephone systems. The argument that so much copper cable has already been committed to the ground that radiophones will not supplant fixed wireline telephones does not hold true in the many developing countries of Asia. So if wireless does become more of a substitute and less of a complement of wireline — here cost will be the important consideration — Asia could be the experimental ground for it.

Huber, Kellogg and Thorne (1993, pp.19–20) summarize the consequence of the two changes as follows:

> Local radio networks will not eliminate copper, but they will overlay it quickly enough. And both radio and microelectronics are pushing the local network in precisely the same direction — away from long, expensive, star-like pyramids with utility cost structures, towards short, cheap, ring-like geosodics, in which competition prevails.

This may be Asia's future, and Hong Kong in particular is moving substantially towards it, but the present is still a period of transition from the state monopoly utility towards more open markets in which value-added services, like paging, mobile cellular telephony, store-and-forward fax and data, online information services, are establishing themselves as 'supermarket' commodities, but basic voice services remain the domain of the incumbent national telephone company. A key question therefore arises: is there a model of the transition which can be applied to Asia to predict not just the long-term outcome — Huber et al provide that — but the stages and speed of the transition?

Noam (1994) has proposed an 'evolutionary' model of network development which goes through six stages. In stage one the network is small and offers little utility to new subscribers, needing a subsidy to fund it. Stage two sees average costs falling as economies of scale kick in and the utility of the network to subscribers grows. Stage three begins when the trend in average costs reverses, and the additional utility brought to existing subscribers by the marginal subscriber diminishes.[7] The 'private optimum' point of network growth has been reached and the phase of 'self-sustained growth' closes. Stage four sees the network continue to expand, driven by the social goal of universal service, but increasingly this requires an allocation of resources and a tariffing policy which is redistributive. Noam calls this phase 'entitlement growth.' Stage five sees average costs rise above the value brought to a sub-set of existing subscribers by further network expansion. This sub-set then has an incentive to opt out of the existing public network and into stand-alone private networks. Stand-alone networks have limited utility unless they can interconnect with the public network, so a key regulatory issue here arises over the right to interconnect, and at what cost. Stage six sees the universal service obligation requiring an increasing subsidy, and probably coming under attack from the user groups who pay for the subsidy.

This model is useful as a benchmark, and clearly reflects a historical reality, but not contemporary reality in developing Asian economies. There is a twofold reason for this. First, technology has jumped radically, rendering a purely evolutionary approach inadequate. Second, economic development in Asian countries is combined and uneven, as discussed above. As a result, some regions and cities are racing to install leading-edge technologies in their telecommunications networks, opening them to private network interconnection, doing all the things they are supposed not to do until the country has advanced to stage five. Yet the country as a whole remains at stage one or two, with a target of reaching stage three by the end of the century. The Noam model as a reference model is therefore useful precisely because Asia represents a departure from it.

Policy, Planning and Development

The most important policy change in Asia over the past decade has been the priority assigned to telecommunications in national economic and social planning. The major driver has been the need to modernize in the face of growing competition from the world market, and here policy has had two aspects: proactive and reactive.

The most proactive have been the two island economies of Hong Kong and Singapore, the former by liberalizing its markets and the latter by promoting an island-wide information infrastructure; and they are driven by the competition between them to be the premier telecommunications hub of the region. In contrast, reactions have been triggered throughout the region by the politics of world trade negotiations, and especially GATS — the General Agreement on Trade in Services — with Asian governments reluctant to open their economies, and their telecommunications markets, too far too fast. At the same time Asian governments are well aware that the most successful models of self-sustained growth in the region have been those of an export-creation rather than an import-substitution type (World Bank, 1993) and they have been opening their markets steadily by degrees. The telecommunications industry has played a key role in these growth strategies over the past decade.

These themes are explored in greater detail by John Ure in Chapters 2 and 3. Chapter 2 looks at China's enormously ambitious network building programme, and the implications for the telecommunications sector as state enterprise reform opens the door to domestic competition. The chapter also examines in detail changes in policy and regulation in Hong Kong, Singapore, South Korea and Taiwan. Chapter 3 looks at the economies of the developing countries of ASEAN and Indochina.

A further key driver in the reform process during the 1980s was the fiscal crisis of the state. In some cases, for example in Malaysia, this was the decisive factor in the timing of the restructuring of the industry, shifting the management and operation of the public switched telephone network (PSTN) from a department of government to a self-financing corporate enterprise. As a further step the partial privatization of these corporations has taken place, for example in Indonesia, Malaysia, Singapore, as a prelude to the opening of the market to competitive entry. This process has been problematic for governments because more often than not the net revenues of the national telecommunications operator have been claimed by the state treasury (usually to the Ministry of Finance) to supplement the national budget.[8] Hard foreign currency earnings from international traffic have proved particularly attractive to governments.

The fact is that governments can only meet their plan targets for raising teledensity — telephone mainlines per 100 inhabitants — by opening the sector to private sources of capital. In Chapter 4, Andrew Harrington takes this fact as his point of departure. He surveys these developments, identifying the variety of local forms this process has taken, and argues that as more countries open their telecommunications markets, a competition for capital is emerging. Not only do countries have to compete for funds from capital markets and private investors, but telecommunications as a sector increasingly has to compete against other infrastructure sectors, such as power and transportation.

The 'other services' the Maitland Report refers to are those which are frequently categorized as 'value-added' or 'enhanced' services, such as data network traffic management, store-and-forward voice and facsimile messaging, roaming services using cellular mobile telephones and pagers across country borders, and so on. Increasingly these services are internationally tradable — as international value-added services (IVANS). New technology facilitates this trade. The capacity that submarine optical fibre cables provide for high-speed data communications, for example, gives rise to worldwide corporate computer networking. The growth in satellite transponder capacity across the Asia region is opening the way for many new businesses, such as cross-border multi-facsimile transmission, cross-border mobile voice and data roaming services, regional and international information services. These regional trends are analysed in detail by Peter Lovelock in Chapter 5.

The micro-electronics revolution has brought about a convergence of technologies across three industries, broadcasting (and narrowcasting), computing, and telecommunications, each with entirely different histories of ownership, control and regulation. This plays havoc with existing regulations, especially as they each have technological possibilities and commercial interests to encroach into the others' markets, for example telephone traffic can be carried by cable TV and computer networks, while telephone networks can offer video-on-demand and interactive information services. Each is exploring a range of products and services known as multi-media. Within domestic markets, cable transmissions will provide the most satisfactory means of delivery of multi-media services, such as video-on-demand, owing to the bandwidths required to carry moving image as well as voice and data on an interactive basis. But across the region media transmissions are proliferating in parallel with the number of satellites available. For many Asian governments this raises the perceived threat of trans-border encroachment of non-Asian (or non-national) culture, values and ideologies. In many cases it has led to governments trying to enforce a ban on satellite dishes to prevent reception of foreign broadcasts. In Chapter 6, Peter Lovelock and Susan Schoenfeld explore these issues as they surface in Southeast and East Asia.

Asia now boasts five of the world's top ten telecommunications equipment markets, and in Chapter 7 Nick Ingelbrecht examines how equipment suppliers, who previously enjoyed protected markets selling to local state monopolies, are having to change their strategies in line with the liberalization of Asian telecommunications. The aspirations of Asian states to develop their own manufacturing capabilities in the field of micro-electronic parts and equipment pose a challenge to world manufactures. On the one hand, Asian states want to negotiate the terms of entry of multinationals into their markets in return for the transfer of their technology and skills to assist the development of local industry, while multinationals are reluctant to give away their firm specific advantages and possibly sow the seeds of their own destruction. On the other hand, there is increasing competition between multinational suppliers, with research, development and product cycles shortening, and this places an urgency upon their entry into Asian markets and their ability to reach local agreements to joint venture, or license local production or establish a chain of local component suppliers to support a wholly owned facility. In consequence, as Nick Ingelbrecht suggests, a reversal of the flows of supply and demand of the past is occurring.

Events in the Asian telecommunications industry are moving so fast that many of the details contained in the following chapters will become yesterday's news quite quickly. For that reason these chapters have been written to provide deeper insights into the workings of these markets, and to provide a framework of analysis which will remain relevant for a good time to come. The test of that will be to see, in a few years, how well developments in the later 1990s can be understood in terms of the structural and market dynamics presented in this volume.

1. The decision not to include Japan was governed by the focus on developing economies, and because a proper discussion of Japan's policy, network and services development would require a major addition to a book that is already substantial in content. The decision to exclude the Indian sub-continent was based partly on the same reason, and partly upon the fact that, at the time of writing, the editor has little detailed knowledge of Bangladesh, Bhutan, India, Nepal, Pakistan and Sri Lanka.
2. Asia as a whole accounts for 58 per cent of the world's 5.5 billion population, and claims 21 per cent of main telephone lines, but Japan alone accounts for 10 per cent.
3. The ratio of telephone sets to telephone mainlines varies from country to country, with 1.5 probably about average. See ITU (1993).
4. This was interpreted to mean within a day's walk.
5. Of the ASEAN states only Malaysia has a telephone mainline density (telephone mainlines per 100 inhabitants or 'teledensity') higher than the world average — which in 1992 was 10.49 and Malaysia had 11.13 — or indeed higher than the average for lower-middle income countries — in 1992 this was 7.05. The world average for high income countries was 49.14.
6. An Italian engineer, Guglielmo Marconi (1874-1937) pioneered the development of radio. By 1895 he had accomplished ship-to-shore communications of over one mile, and in 1900 he succeeded to establish direct wireless communications across 1,700 miles of the Atlantic Ocean. In 1909 he was the joint winner of the Nobel Prize for Physics.
7. It is not immediately obvious why average costs start to rise in Noam's model. One reason could be the universal service obligation to meet the demand of subscribers in remote areas, that is, rising costs at the external margin. But in practice this seems to ignore rapid (Schumpeterian) technological change (e.g. satellite and microwave technology) which substantially reduces real costs. Under the old electro-mechanical and semi-electronic switching technologies, and copper cable transmission networks, telecommunications engineers were divided between those who believed average costs would rise as the service area became more dense, and those who saw constant costs at the intensive margin. (See Littlechild, 1979, p.49). Noam's model seems to abstract from technological change entirely.
8. The World Bank found that, during the 1980s, 32 per cent of the net revenues of state-run telecommunications carriers in Asia were transferred back to the state.

Annex to Chapter 1

The following table illustrates the quickening pace of growth of the public switched telephone network (PSTN) in the developing economies of Southeast and East Asia over the past decade.

Table 1.1
Telecommunications Network Growth in Southeast and East Asia

Country	CAGR (1980–92) Percentage	CAGR (1987–92) Percentage	Teledensity (1992)	Mainlines millions (1992)
China	14.9	20.4	0.98	17.33[i]
Hong Kong	6	7	48.62	2.8[i]
Singapore	6	6	41.6	1.17[i]
South Korea	12[a]	12[b]	36.34	15.9
Taiwan	8[c]	8[d]	35.75	7.7[i]
Indonesia	11	14	0.81	1.5
Malaysia	14	21	11.13	2.1
Philippines	6	9[e]	1.03	0.9[i]
Thailand	14	17	3.10	1.9
Burma	7.3[f]	2.3[g]	0.18	0.08
Cambodia	–	–	0.06	0.005
Laos	2.2[f]	1.3[h]	0.15	0.007
Vietnam	12.7[f]	13.6[h]	0.29	0.2

a. 1982–1993; b. 1987–1993; c. 1985–1992; d. 1986–1992; e. 1985–1992; f. 1983–1992; g. 1989–1992; h. 1988–1992; i. 1993.

Source: National telephone administrations; *World Telecommunications Development Report* (ITU, 1994).

The cumulative annual growth rate (CAGR) is measured as the annual rate of increase in main exchange lines directly connected to the subscriber's premises equipment. (Note that the figures for China relate to total local exchange capacity.) The CAGR of the poorest countries increased noticeably over period 1987–92, Burma and Laos being exceptions. In the middle and high income 'dragon' economies — Hong Kong, Singapore, South Korea, Taiwan — growth remained constant.

Teledensity is the ITU's term for the number of main telephone lines per 100 inhabitants. According to the *World Telecommunications Development Report* (ITU, 1994) the teledensity average of countries in 1992 was:

Low Income	0.80
Lower Middle Income	7.05
Upper Middle Income	13.49
High Income	49.14

These figures place the dragon economies among the high-income countries, but only Malaysia among the middle-income countries. Mainlines refers to exchange lines directly connected to the subscriber's premises equipment. As the CAGRs suggest, these are rising rapidly, not more so than in China where the Ministry of Posts and Telecommunications (MPT) announced that by the end of 1994 the number of telephone subscribers had doubled to 37.43 million. (China Daily, 13 January 1995, p.1). The MPT also announced there were 1.57 million mobile telephone subscribers.

Telecommunications in China and the Four Dragons

John Ure*

China tends to dominate this chapter, not because the four dragons are closely related communities with China as the uniting factor, but because of China's size and potential for growth.[1]

By 2000, China aims to have an exchange capacity of around 140 million main telephone lines. Today only the USA has more, at nearly 150 million. By then also the British colony of Hong Kong and the Portuguese colony of Macau will have been returned to mainland Chinese sovereignty. Fifty per cent of Hong Kong's international traffic is with the mainland, and 30 per cent of Macau's, with another 60 per cent to Hong Kong. The future of Taiwan remains uncertain, but since the lifting of the island's state of emergency in 1988 telecommunications traffic between Taiwan and the mainland has risen to around 14 per cent of Taiwan's total outbound traffic, although owing to the restrictions Taiwan places upon direct communications with mainland China, this traffic is routed through Hong Kong, Japan or the USA.

Singapore is less oriented towards China and Hong Kong, at least in terms of trade and traffic. (See Chapter 5 for a detailed analysis.) In 1992, only 8.3 per cent of Singapore's outbound traffic went to Hong Kong, and

---

*   I am indebted to many people in the telecommunications industry throughout the region, too numerous to name, for their generous assistance with this and the following chapter. I would especially like to thank Nick Ingelbrecht for his patient comments on an earlier draft. I remain solely responsible for the contents and views of Chapters 2 and 3.

China did not even figure in the top ten most called countries. But South Korea is rather more involved, with Hong Kong the third most called country, although accounting for only 4.8 per cent of outbound traffic, and China the fifth most-called country, accounting for 3.5 per cent. (Again, see Chapter 5 for more details.)

What all these economies have in common, including mainland China's coastal industrial economies around cities such as Beijing, Tianjin, Shanghai and Guangzhou, is their openness to the world economy, which is driving the demand for highly sophisticated telecommunications network facilities and services. The need to manage or promote a transition to a higher value-added industrial structure has been recognized in each of these economies, including sectors of the coastal economies of China, and since the early 1980s telecommunications development has been designated a priority within science and technology funding and industrial policy. Examining the transition in terms of policy is the theme of this chapter.

China

The Legacy

For historical reasons, the People's Republic of China suffers from a chronic underdevelopment of its telecommunications network, with less than two mainlines per 100 inhabitants. The Stalinist model of the state, introduced after the coming to power of the Chinese Communist Party in 1949, discouraged private access to the means of communication. As a project, the democratic revolution, essential in the political and social ideas of Karl Marx, was simply abandoned. By rapidly absorbing the remnants of civil society into the administrative and political structures of the state, Communist rule also reduced the sphere of demand for communications that naturally arises from personal commercial and social activity.

Following the death of Mao Zedong in 1976, China's new patriarch, Deng Xiaoping, returned China's economy to a course abandoned in the mid-1950s. From 1979 China pursued an open door policy, beginning with special economic zones designed to attract foreign investment, technology transfer and promote exports. This was followed by market and property ownership reforms in agriculture, commerce and eventually in industrial and state enterprises. Long, renewable and transferable leases have effectively re-introduced private property to the PRC. And the 'iron rice bowl' which guaranteed workers lifetime employment, and the provision of free or low-cost housing, education, health care and other welfare benefits by the work unit, is being broken.[2]

Following Deng Xiaoping's reforms, the 'socialist market economy' is the new guiding principle of China, but this creates a contradiction within a Stalinist system. Under this model the party and state structure parallel each other from the highest levels, the Politburo and the State Council, down to the lowest party cell and work unit, whereas a socialist market economy implies greater autonomy of the enterprise. Enterprise reform cuts against the traditional role of the party cadre as commissar guiding the industrial and commercial activities of what used to be organs of the state.[3]

The contradiction between civil reform and political control now runs through all aspects of China's economic and industrial development, including telecommunications and related information technology sectors on the one hand, and the media-related issues, such as broadcasting, on the other. (See Chapter 6.) Put simply, in an economy that is becoming liberalized and market-oriented, people need access to the means of communication and information technologies, such as computer networks, databases, international E-mail, and the freedom to use them as, when and how they are required. Currently, the shortage issue creates problems of access, and the control issue creates problems of usage.[4]

Recent Reforms

By the beginning of 1994, China had a public network exchange capacity of 42 million lines, but only 17 million subscribers, of whom 14 million were hooked up directly to the public switched telephone network, the PSTN, and 3 million were on small independent rural networks operated by Township and Village Enterprises (TVEs) which interconnect with the PSTN at the nearest point. The local PSTN is operated by the Posts and Telephone Administrations (PTAs) and under the PTAs each provincial city has its own Posts and Telephone Bureau (PTB), and each of these is responsible for groups of six to eight county-level offices. In total China has around 350 PTBs and over 2000 county-level offices.

The PTAs are responsible to the Directorate-General of Telecommunications (DGT) under the Ministry of Posts and Telecommunications (MPT) in Beijing. The DGT is directly responsible for the national trunk and international network, while responsibility for the local networks rests with the PTAs. However, a lot of the funding and support also comes from the local governments (see below) of the four metropolitan areas (Beijing, Guangzhou, Shanghai and Tianjin), twenty-one provinces, and five autonomous regions, because they regard telecommunications as an asset in their competition to attract domestic and overseas investment to their region. The urgency accorded to network development is underscored by the absymally low successful call ratios (SCRs) during busy hours (which are frequently less than 60 per cent, often less than 30 per cent) and by waiting lists which have been growing annually by an average of 30 per cent, reaching 1.6 million by 1992 according to the ITU (1994).[5]

Recognition of the importance of telecommunications to economic growth and regional competitive advantage[6] is driving changes in China's telecommunications policy. In 1993 and 1994 the State Council, the highest level Government body, approved the setting up of two corporations, Jitong and Liantong, or United Communications Corp (Unicom),[7] both initiated by the Ministry of Electronic Industries (MEI), to provide alternative, or supplementary, networks. Jitong is to build out at least three of the so-called Golden Projects[8] by constructing satellite, microwave and cable connections to form medium-speed national information networks. According to MPT Minister Wu, this project, known as the Golden Bridge, will provide an information resources network linking databases for private end-users, whereas the MPT will be providing an 'online' network of information sources for public use. (*Eastern Express*, 14 April 1994, Hong Kong). Liantong is authorized to build and operate both fixed-line and radio-based local and trunk telecommunications networks. Many ministries and state enterprises run their own networks,[9] and China's chronic national shortage is excuse enough for these ministries to lobby government to allow them to enter the market. (See Ure, 1994a.)

At local levels, radio communications are widely used as vehicles of market entry. Radio spectrum comes under the authority of the State Radio Regulation Commission (SRRC) and its provincial branches,[10] but many ministries and local bodies, such as the regional commands of the People's Liberation Army (PLA), local branches of the State Security Bureau, which is responsible for police and emergency services, and numerous state enterprises control spectrum for the operation of mobile phone, pager, trunked radio and walkie-talkie systems. These networks are supposed to be non-public,[11] but interconnection and revenue-sharing agreements with local PTBs often blur the distinction between private and public. This offers opportunities for non-mainland companies, especially from Hong Kong and Taiwan, to enter the China market through the backdoor by providing capital, network build-out and management consultancy, all of which is repaid, in effect, from network revenues. (See below.)

A reorganization of the MPT in 1994 separated off the national network operator, the DGT, as a state enterprise responsible for its own finances, from the MPT as the national regulator funded directly from the state budget. In theory the MPT should become an impartial overseer of Jitong and Liantong, and other new entrants as they arise, guaranteeing fair interconnection and revenue-sharing arrangements between networks. In practice the rivalry between the MPT and MEI,[12] which extends into the field of telecommunications equipment manufacturing, is sharp, and to enforce at least the appearance of peace a Unified Conference has been established under the leadership of Vice-Premier Zhou Jiahua to whom these ministries are responsible. The reorganization has been assisted by the World Bank which is funding consultancies looking into modern telecommunications accounting procedures and tariffing polices.

So China's telecommunications network sector is undergoing radical policy shifts. One of the ironies of China's legacy is that alternative networks were created by many different ministries, and, as Zita (1994) points out, as economic decentralization takes place these ministries have an incentive to upgrade and improve their management information systems. This makes them more able to offer network capacity to alternative public service providers, such as Jitong and Liantong. Further, State Council Order (1993) 55 announced the liberalization of the following sectors subject to receiving a licence: radio paging, non-public cellular mobile telephone (450 MHz and 800 MHz), Vsats; and, subject to a declaration, the following: telephone and computer messaging services, E-mail, EDI, and videotex. (See *China Telecommunications Construction*, October 1993, pp. 6–7; and Beijing *Economic Daily*, 20 October 1993, pp.3–4). Non-MPT network direct investment and service operation is allowed in these areas, but foreign direct investment and management has remained prohibited. VANS, that is a network over which value-added services are offered, can only be operated by Chinese national companies, but VAS, that is value-added services offered over a network operated by a Chinese company, are open to foreign companies. IVANS, that is international valued-added services operated by foreign companies also remain prohibited, but concessions are likely (see *Communications Week International* 12 December 1994, p.4) given China's commitment to membership of the World Trade Organization (WTO). For insights into China's telecommunications policy, see Lee (1995), and especially the China Special Issue of *Telecommunications Policy*, edited by Mueller (1994).

Telecommunications and Economic Reform

In centrally administered China the role assigned to any major industry is multifaceted. Telecommunications are part of the national security network; they are part of the drive towards national economic modernization; they are part of Beijing's central need to keep in touch with, and control over, the outlying provinces. Since the open door policy, and especially since Deng's 'tour of inspection' to the south in 1992 — as part of his campaign to wrestle the initiative for reform back from the conservatives[13] — during which Deng called for faster economic growth, China's economy has become more vulnerable to economic swings and trade imbalances. The domestic supply of money and credit threatens to get out of control as state enterprises queue up for loans to cover their debts, or just to speculate, with the resulting inflation sparking social unrest. Under these circumstances, Beijing's central planners — conservatives and reformists alike — support the need for modern information technology networks to make the planning process, or at least the control process, a real one.

The framework for the long-term role of telecommunications was also laid down in the mid-1980s in line with the renewed emphasis that Deng placed upon science and technology in the modernization of China.[14] A Leading Group for the Revitalization of the Electronics Industry, established in 1984 and led by then Vice-Premier Li Peng, took responsibility for planning telecommunication sector growth as part of China's drive into the electronic age. (See Zita, 1987.) Just as South Korea and Taiwan have used PSTN procurement policy to stimulate local manufacturing, and Singapore has used network development to encourage the dispersion and use of information technology, so in China telecommunications is used as an instrument of industrial policy, a vehicle to stimulate domestic equipment and components manufacturing. This role has grown in recent years as China insists upon foreign suppliers of telecommunications equipment entering into technology transfer agreements, usually through joint venture manufacturing, but full foreign ownership is also permitted. In 1985, under this stimulus, the MPT submitted to the State Planning Commission a fifteen-year forecast which appeared as the report *China to the Year 2000*.

Then the target for 2000 was 33.6 million telephones (telephones normally outnumber mainlines) and a national teledensity of 2.8 mainlines per 100 inhabitants. The target for 1990 was around 9 million lines. Almost immediately steps were taken to accelerate network development following the endorsement by the Communist Party Central Committee and the State Council in 1986 of Project 863 (drafted in March 1986, hence its name) which listed communications and information technology among seven priority areas for development. The thrust of what is otherwise called the Chinese High-Technology Research and Development Programme (see *Telecommunications Development Asia-Pacific*, September 1994, pp.67–70) is basic and applied research and development,[15] but, starting 1988, economic assistance was given to the MPT in the form of the 'three 90 per cents': 90 per cent of central government loans need not be paid back, PTAs could keep 90 per cent of their taxable profits, and the MPT could retain 90 per cent of its foreign currency earnings from incoming international traffic.[16]

The 1988 financial reforms were accompanied by a reorganization within the MPT as a step towards giving the PTAs a greater degree of autonomy in procurement policy, in investment decisions, in local planning and accountability for financial performance. For example, staff are paid bonuses according to traffic successfully carried, and local levies can be added on to the tariff levels set by the MPT, and approved by the Price Boards, to raise additional investment finance. (See Bruce and Cunard, 1994.) The net result has been faster network growth, and rapid innovation in areas such as cellular mobile and paging services. By 1990, mainlines in China had reached over 12 million, still desperately low by international standards, just 1 per 100 inhabitants, but one-third above target. Table 2.1 provides a detailed breakdown of telephones by province for the year 1992.

**Table 2.1
China's Telephones by Province 1992**

	Total (000s)	% of Rural	Phones /100		Total (000s)	% of Rural	Phones /100
North				**South**			
Beijing	1237	2	11.2	Guangdong	2631	33	4.0
I' Mongolia	364	14	1.6	Guangxi	309	18	0.7
Hebei	791	14	1.4	Hainan	117	12	1.7
Shanxi	426	11	1.4	Henan	640	12	0.7
Tianjin	429	7	4.6	Hubei	611	19	1.0
Hunan	532	25	0.8				
North-East				**West**			
Heilongjiang	746	11	2.0	Guizhou	169	14	0.5
Jilin	603	18	2.3	Sichuan	804	14	0.7
Liaoning	1067	13	2.6	Tibet	19	2	0.8
				Yunnan	257	257	0.6
East				**North-West**			
Anhui	457	15	0.7	Gansu	263	9	1.1
Fujian	564	16	1.8	Ningxia	70	7	1.4
Jiangsu	1505	26	2.1	Qinghai	72	4	1.5
Jiangxi	332	19	0.8	Shaanxi	401	12	1.1
Shandong	1036	21	1.2	Xinjiang	245	14	1.5
Shanghai	1157	11	8.6				
Total	18888	18	1.6				

Source: *China Statistical Yearbook 1993*, Beijing.

Following Deng's 1992 'tour of inspection', targets were raised, and then raised again, which underscores the point that these are top-down targets, based upon decisions taken in Beijing, and upon perceptions of national priorities. The target for 2000 was brought forward and then raised to 50 million lines by 1995, and 100 million by 2000. During 1994, the 2000 target was raised again to 114 million lines to be provided by the MPT. As a result of the 1993 and 1994 reforms (see above) which will permit non-MPT networks to compete, a total of 140 million lines should be available by 2000, quadrupling the original target set in 1985. If successful, nationally China will still only have around 8 lines per 100 inhabitants, but in the major cities the target is closer to 40.

Investment in Telecommunications

Meeting the targets may be less of a problem than at first appears. Domestic production of switches, for example, including the capacity of the joint ventures China has with Alcatel, NEC and Seimens, and two planned with AT&T and Northern Telecom, is forecast to reach 10 to 12 million a year by 1997. Financing may also be less of a problem. The ITU (1993) rule-of-thumb of US$1,500 per line seems far too high for China. Xu (1994), using World Bank sources, suggests US$800 is more likely, but there are reasons for thinking this is also too high. One of the effects of the 1980s reforms has been for the PTAs to develop much closer ties with local government than with Beijing, not least because local governments view telecommunications as an important asset in the competition between provinces and cities for foreign and domestic investment. For this reason provincial governments often provide cables for the local loop free to their PTBs, and the local garrison of the PLA also provides free labour for the construction of ducts. In this way there are many subsidies that go to the local networks.[17] Switches are manufactured in China at less than US$80 per circuit, and another hidden subsidy is the research and development that goes into the software, often carried out in Ministry of Defence research laboratories.

On the other side of the equation, sources of finance are still quite plentiful. Domestic capital flows freely into the telecommunications sector, often in the form of leaseback arrangements with the PTAs. The high registration and connection charges, often around US$500, and local surcharges on tariffs more than finance network installation costs. During the past decade state funding has been only one-third of total funding; of the rest half comes from PTA sources, and the rest from loans, domestic funds and vendor credits. As Liang and Zhu (1994, p.78) point out, 'Foreign capital has remained negligible.' The opportunities to increase the generation of revenue grows as the network grows, and tariff restructuring could further add to revenue growth. (See Ure, 1994b.)

One area of great profitability for the PTAs has been mobile telephones. Handset prices and registration fees have been as high as US$3,000 or more, yet in the more prosperous coastal regions there has been no shortage of demand. Part of this demand is for substitute access to the PSTN, but part is also for the functionality, and in many cases the fees are paid for by companies rather than individuals. PTAs can recover their capital outlay within one or two years, and pay foreign systems suppliers, such as Ericsson and Motorola (the two leading suppliers of analogue handsets) cash for turnkey networks. In the late 1980s, the MPT's original market projection for 1995 was 30,000 subscribers. By the end of 1994, there were close on one million, half of them in Guangdong province to the north of Hong Kong, and forecasts for 2000 are close to 10 million. CT2 networks have also proved popular in several cities, and pagers, which already numbered close to 10 million by the end of 1994 compared with under 500,000 in 1990, are projected to reach 25 million by 2000. These are all sources of rapid revenue and high returns, so much so that the MPT in 1994 instructed PTAs to reduce handset prices, raising suspicions that this was a move to undermine a potential revenue base for new entrants. Liantong, for example, has been awarded by the SRRC 900 MHz frequencies to build and operate digital GSM cellular networks in the major cities.

A joint decree (Order 128) issued by the Central Military Command (CMC) and the State Council in 1993 (see *People's Daily*, 11 September) clearly recognized the growing commercial value of radio frequency. In an apparent compromise between the Ministry of Defence and the MPT, they assigned the rights to allocate civilian and military frequencies for commercial purposes to the SRRC and PLA respectively. This strengthens the military's hand to use freed-up military spectrum in joint ventures with whoever they choose as partners. (But also see Chapter 7.) Some of those partners are already Hong Kong, Taiwanese or other Sino-Asian companies, providing capital, equipment and management expertise.

For an international telecommunications company, entering into arrangements with non-MPT entities is still a risky business, from the viewpoint of both commercial law,[18] which is hardly developed in China, and of business planning, of which there is little experience in China. Local Asian companies are in a more flexible position. They tend to be smaller, frequently family-controlled, and less restricted by the concerns of stockholders and corporate lawyers when entering into business deals in China. Where they are run by ethnic Sino-Asians, such as Thais or Indonesians of Chinese descent, the ancestral connections with particular provinces in China can figure prominently in investment ties. Throughout Asia the accumulated savings of ethnic Chinese business families (we could add ethnic Indian Sub-Continent business families also) are an important source of domestic investment, which is perhaps part of the cause rather than effect of the relatively low ratio of foreign direct investment to fixed investment in Asian economies. (See Ramstetter, 1993.)

Asian capital investment in China's telecommunications sector is prepared to enter a grey area of joint ventures with local partners who alone can, according to Chinese regulation, own and operate the networks. While foreign direct investment in networks remains banned, there are signs that foreign companies are exploring ways either to gain equity-type returns[19] or to structure their equipment and building contract loans on the basis of an initially low payback, with the understanding that if and when the rules towards foreign direct investment are changed these loans may convert to equity.

The Information Economy

While residential customers want the basic service, China's drive to attract foreign business investment and encourage local economic development, will create a demand for information and data networks. A national grid of optical fibre cables is being built to link up all China's major cities and this, together with the rapid expansion of China's satellite programme, should aid an equally rapid expansion of China's infant information technology industry. In 1994, there were only 800 acknowledged databases in China, mostly not online,[20] and official estimates placed the installed base of PCs at less than 1.5 million (see *China Daily*, 12 July 1994), but the State Information Centre, under the State Planning Commission, is promoting the need for the development of the information economy (see Wu, 1994), while many enterprises are exploring the market.[21] Chinapac, a packet-switched data network (PSDN) has been extended nationwide, riding on the back of a public digital data network (DDN) that was cut over in 1994 (see *China Daily*, 8 November). For a good insight into computer networking in China, see Zheng, 1994.

China is also jumping on the information superhighway bandwagon with a high-level committee studying the feasibility of a national highspeed broadband network. Guangzhou, Shanghai and Beijing are already putting into place, or planning, SDH (Synchronous Digital Hierarchy) highspeed data transmission networks, and in Guangdong the American company SCM/Brooks is helping to finance the Huamei company, a joint venture with the PLA's Galaxy New Technology Company, to build an ATM (Asynchronous Transfer Mode) broadband switching system with AT&T equipment. This promises to bring China into the multi-media age, although cooperation and coordination between telecommunications and cable TV networks, run by the Ministry of Radio, Film and Television (MRFT) remains uncertain. (See Chapter 6.) The network will provide services to hotels and exhibition and trade centres in Guangzhou, and appears to involve some degree of foreign direct investment, which may be an interesting case of regulations being flexible where convergent technologies and multi-media services are contemplated.

Conclusion

China has been without a telecommunications law governing the industry since 1949, although according to Minister Wu (*China Telecommunications Construction*, v.6.5, October 1994, p.11) the drafting of such a law is under preparation for submission to the recently revitalized National People's Congress. This vacuum left the industry without direction until the early 1980s when the sector was recognized as of strategic importance. Since then it has been promoted through basic and applied research and development, and through financial incentives and organizational reforms. The national targets for the year 2000 and beyond have placed telecommunications at the top of China's industrial agenda, alongside power generation and transportation.

The promotion of an information economy is part and parcel of China's drive to modernize, and is seen as an essential element in the ability of the government to manage the economy, and to govern. The sector is undergoing restructuring as part of the state enterprise reform process, and this has interesting implications for its future shape. The DGT has, in effect, been corporatized, and its first step has been to form separate operating companies for mobile, data and value-added services. In the spirit of the 'socialist market economy', it must run the national and international networks and services on a commercial basis.[22] This means becoming self-financing, tariff rebalancing, more customer-oriented, more innovative in its services, and sourcing capital in the most effective means possible.

But if the PTAs are also to follow the path to a 'socialist market economy', they too will start asserting the right to source their capital from the most cost-effective sources, and partnering with companies that can bring the technology. It is therefore interesting to speculate if and when PTAs and local governments will demand the right to experiment with Build-Operate-Transfer arrangements in local telecommunications networks, or other forms of equity holdings. In Minister Wu's statement, the PTAs are also extolled to 'seize opportunities, carry reforms to more depth, open wider to the outside world, promote further development, and keep progress steady'. These words are loaded with unexplained meanings.

One pressure on the PTAs to go in this direction will be competition at the local level from new domestic entrants. For example, besides Jitong and Liantong and the PLA, there is the Ministry of Radio, Film and Television which is building local cable TV networks in all the major cities. These could carry telecommunications services. (See Chapter 6 for details.) There are also national organizations like the Ministry of Railways which is upgrading its own network, and the Bank of China which is constructing a data network of its own. These and other networks could, at some future stage, offer capacity to customers in a country chronically short of telecommunications facilities and services. Already privately run networks are being set up at provincial level, such as the Shuntung Wireless Telecommunications Company in Anhui Province. (See *China Daily*, 16 November 1994.) Domestic competition seems to be arising in China's telecommunications markets in a way that reflects provincial rather than central decision-making, and this is not how the MPT envisages things, since models of provincial operating companies which would do away with the MPT altogether have never found favour.

Finally, the financial problems of funding China's enormously ambitious development plans can be eased by sensible tariff reforms, while growing network revenues will not just tap domestic funds, but will make the raising of foreign loans through investment funds and overseas capital markets relatively easy. By itself, finance is unlikely to be the critical factor. (But see Chapter 4 for an alternative view.) What China lacks most of all is expertise in making the transition from a bureaucratically administered basic service to a responsive commercial organization capable of managing the services which lie at the heart of a modern telecommunications network, which can provide the platform for an information technology sector.

The pressure for reform will come from a growing number of customers, as is already evident from the support given to Jitong and Liantong. But it will come even more powerfully from a growing number of PTAs which see themselves under pressure, and demand of Beijing the right to enter into joint ventures with foreign partners who can bring not just the finance but the expertise as well. The interesting question is therefore not when will China allow foreign participation in networks, but how. China needs to serve its rural areas, and transfer technology and experience from the developed world. The way China finds solutions which meet these economic and social goals will have implications far beyond its borders.

Hong Kong

Among the four dragons, Hong Kong vies with Singapore for leadership in telecommunications services. Against Singapore's teledensity of 44 telephone mainlines per 100 inhabitants, Hong Kong boasts 3 million mainlines connected, or one for every two persons. The switched network is entirely digital and the mesh connecting telephone exchanges is all optical fibre, as are many of the cables servicing Hong Kong's central business district. Because there are no local timed call charges the use of facsimile is also encouraged, and there are now a quarter of a million separate fax lines. And on many international routes, for example to Japan, fax traffic now outstrips voice traffic.

Like Singapore, Hong Kong thrives as a communications and trading hub, especially for the burgeoning economy of southern China's Guangdong Province. Unlike Singapore, most of Hong Kong's manufacturing base has migrated, moving across the border to exploit cheap labour and land prices. The balance of employment in Hong Kong has consequently shifted to the service sectors, especially wholesale and retail, restaurants and hotels, business and personal services, and transport and communications. These sectors are substantial users of both fixed-wire and mobile communications.

Hong Kong's population of 6 million is twice that of Singapore, as is its territory of 1,000 square kilometres. Hong Kong's domestic economy is around three times larger, but both depend heavily upon trade.[23] Hong Kong's industrial structure is overwhelmingly characterized by small and medium-sized enterprises employing less than 200 staff,[24] and which do not generate very high levels of domestic demand for sophisticated telecommunications services, but the economy is dominated and driven by sectors that do. Hong Kong is a centre of international banking and finance, sea and air transportation, hotels and tourism, trade and international conferences, and the regional hub and headquarters of around 400 multinational companies. See Ure and Chen, 1993. As Petrazzini (1994) points out, approximately 70 per cent of all US firms with Asian headquarters centralize their operations in Hong Kong.

Since 1925, the domestic PTSN has been the monopoly of HKTC (Hong Kong Telephone Company) under government regulation. The British company, Cable & Wireless, or C&W, provides the international network and voice services through a local subsidiary, originally C&WHK, now HKTI (Hongkong Telecom International), under an exclusive licence which expires in 2006. Early in the 1980s, the government decided to open the customer-premises equipment (CPE) and value-added services markets, including mobile cellular telephones, and required HKTC to adopt separate accounting for competitive services through a subsidiary company, CSL (Customer Services Ltd). These three companies, HKTI, HKTC and CSL, now come under the holding company Hongkong Telecom, formed in 1987 following the take-over of HKTC by Cable & Wireless in 1984.[25] Between two-thirds and three-quarters of C&W's world-wide earnings derive from Hongkong Telecom and in 1993 the two companies formed a joint venture, Great Eastern Communications, which revives an earlier name of C&W, to promote investment in the region, including China.[26]

In contrast to Singapore's proactive state, 'positive non-intervention'[27] is the guiding principle of the Hong Kong government's approach to industrial policy. In practice this means light regulation of public utilities, financial markets, property markets, and so on. As Schiffer (1984) points out, since most land in Hong Kong is government property, rising revenues from land-lease auctions have made a growing contribution to financing social service expenditures, on education, housing and roads especially, while keeping the taxation levels low. So Hong Kong, while committed to a market approach, nevertheless adopts a Keynesian macro-economic pragmatism, underpinned by a Ricardian socialist principle of state control of the scarcity rent.

This approach influences telecommunications and information technology policy. HKTC's monopoly over the domestic PSTN expires in 1995 and three new companies, from seven applications, have been selected for Fixed Telecommunication Network Services (FTNS) licences to offer competitive

PSTN services from that time. Each of the new entrants, Hutchison Communications, New World Telephone, and Wharf Holdings' New T & T are part of the elite among Hong Kong's cash-rich giant property development companies. Hutchison and Wharf are dominant companies and despite early partnerships between the US Baby Bell company Nynex and Wharf, and between Australia's Telstra and Hutchison, both have preferred to go it alone. Only the smaller New World group has partners, which include a 25 per cent stake by another Baby Bell company, US West, and a 5 per cent stake by the Shanghai Long Distance Telephone Company. This choice of new entrants underscores the continuing economic and political clout that local property companies continue to exercise in Hong Kong, and also highlights the fact that in Hong Kong, as in most other Asian economies, local capital is readily available and fully prepared to invest in telecommunications markets.

The risk of such investment in Hong Kong is quite high, especially for an economy where a payback period of five years or less is usually taken for granted. The dominance of Hongkong Telecom will be difficult to challenge, especially in basic voice and fax services, and while the new entrants between them are committed to invest over US$1.2 billion in local networks there are doubts about the size and growth potential of the local value-added services market. For example, neither fax nor mobile data services, both open to competition, have proved profitable to new entrants.The strategy of Hutchison Communications is focused upon the business servicess market by providing a high-speed SDH data transmission network, and also upon leveraging its dominant position in the cellular mobile and paging markets by providing its own fixed-wire connections, and direct delivery to HKTI's international gateway. Under a seperate licence Wharf has a cable TV franchise which will provide direct access to a residential market, and by developing broadband interactive services along an extensive optical fibre backbone, Wharf hopes to penetrate the business sector also through its company New T&T. New World Telephone aims to focus on value-added services, and with a new generation of personal communications, mobile data, and microwave links between computer networks on the horizon it is anticpated that the company will look to new technologies rather than fixed-wire investments to provide the niche market opportunities. But in all cases, if these companies can survive in Hong Kong, they can also look owards the China market in the larger term.

To oversee this new era of competition an independent regulator has been set up known as OFTA, or the Office of the Telecommunications Authority, headed by a Director-General who reports to the Secretary of the Economic Services Branch of government, or ESB. OFTA is modelled along the lines of OFTEL in Britain and AUSTEL in Australia, from where the Director-General was recruited. Following the practice in Britain, HKTC's tariffs are price-capped, in accordance with the formula: Consumer Price Index (CPI) minus X.[28]

In many ways Hong Kong has become a test-bed, and a show-case, for regulation in the region. To promote competitive entry OFTA has taken authority over the allocation of telephone numbers to ensure number portability, both geographically and between operators. Without operator number portability subscribers have been shown to be resistant to shift from the encumbant operator. The precondition for successful entry is the ability to interconnect with the dominant network with guaranteed service and technical quality at a reasonable price. Australia, Britain and the USA have all followed the principle that the new entrants and the encumbant dominant carrier should first try to come to their own commercial agreement on interconnect before the regulator reverted to adjudication. OFTA has followed the same principle. OFTA is also overseeing the revenue-sharing arrangements which will give the new entrants an average of 28 per cent of international revenues from traffic routed through their domestic networks. HKTC continues to receive 40 per cent, the 12 per cent difference being designated the Access Deficit Charge (ADC) to compensate HKTC for its universal service obligation, for example when it supplies the PSTN to Hong Kong's outer islands. This cross-subsidy is likely to be challenged by the entrants when the current scheme comes up for reconsideration in 1996. OFTA is also pioneering in Asia ways to introduce economic pricing into the allocation of one of Hong Kong's scarce resources, radio spectrum.

Both OFTA and the ESB have also made it clear that they intend to interpret HKTI's international license in the most liberal way possible to encourage competition in international services. Satellite down-linking as well as up-linking has been liberalized in contrast to Singapore; closed user group satellite by-pass has been approved, and all international value-added services (IVANS) have been opened. By 1994 there were 30 IVANS operators holding public non-exclusive telecommunications service (PNETS) licences. Domestically, cellular mobile operators can now revenue-share on IDD calls they route directly to HKTI, and from 1995 the same will apply to the three new fixed-wire entrants.

Hong Kong's most spectacular markets are its cellular mobile, CT2 and pager networks. By mid-1994 there were four operators of cellular mobile services with close to 400,000 subscribers. On offer are AMPS and TACS analogue systems, US Digital AMPS and GSM. More cellular and PCN licences are expected to be issued over the next few years, without preconditions on which technologies may be offered. In addition, Hong Kong boosts three CT2 operators serving 165,000 subscribers by mid-1994, and more than 30 paging companies providing a market of over 1.3 million subscribers.Table 2.2 outlines the explosive growth of mobile communications in Hong Kong.

Table 2.2
Mobile Communications in Hong Kong

Date	Pagers Receivers	Analogue Cellular	Digital Cellular	Trunked Radio
December 90	711,428	133,912	N/A	N/A
December 91	880,666	189,664	N/A	3,712
February 92	1,046,384	233,324	N/A	5,344
February 93	1,244,394	253,500	37,343	7,255
August 94	1,327,402	228,793	131,191	7,939

Source: Office of the Telecommunications Authority, Hong Kong.

Despite appearances, Hong Kong's image of being a high technology-using economy is flawed. One estimate suggests that 21 per cent of the population aged 15 and over, or 968,000 people, have a computer at home, and the Telecommunications Research Project of the University of Hong Kong found that 27.5 per cent of households surveyed in December 1994 had a computer, which extrapolates to 435,000 households territory-wide. Only 87,000 of them would have a modem. Another suggests the number of business computers in use by the end of 1994 may be just over half a million, with few of them networked beyond the office.[29] CSL, CompuServe and other online service providers, such as Reuters and Knight-Ridder, are well represented, but mostly for a narrow range of financial and stock market information. See Ure (1995a). E-mail is of growing importance, but still mostly confined to universities and individual enthusiasts using the Internet — several commercial gateway operators are now established, including the part academic and part commercial SuperNet run from the Hong Kong University of Science and Technology — and multinationals using proprietary software. Local databases are very limited. The government-sponsored Trade Development Council is slowly but steadily building a clientele for its trade databases, but a similar effort begun by the General Chamber of Commerce in 1992 has collapsed, while the University of Hong Kong Law Online Database Service has launched a part academic and part commercial online service providing access to Hong Kong and Chinese laws and other Chinese economic and social data. In these latter enterprises commercial does not yet imply profitable, which means that Hong Kong's market-led approach to information technology is being supplemented on a small scale by public institutional funding.

In contrast to Singapore, Hong Kong does not yet have a territory-wide EDI network, although private systems do operate, in the shipping industry for example. The government encouraged a private consortium of major corporations, including Hongkong Telecom, to set up TradeLink to operate a gateway, a Community Electronic Trading Service, to the relevant government departments for applications for textile quotas (Restrained Textiles Export Licences) under the Multi-Fibre Agreement, customs declarations, bills of lading, certificates of origin, and so on. But progress has been painfully slow, and without widely accepted Chinese character software the penetration of EDI into Hong Kong's over 300,000 small and medium-sized firms will not happen.

The debate surrounds the need to speed up the rate of technology diffusion if Hong Kong is to match the other three dragons in their efforts to make the transition to a higher value-added economy. For example, see Kao (1991), King (1992) and Kraemer, Dedrick and Jarman (1994). The government has substantially increased its education spending, including a new University of Science and Technology, and a major expansion in tertiary education, and has merged earlier advisory bodies into an Industry and Technology Development Council (ITDC) supported by a Technology Review Board and a Technology Committee. The Research Grants Council, set up in 1991, allocates and monitors academic research funding, and in another initiative Hong Kong Applied Research and Development Funds Ltd was established to provide matching funds for private companies. The government also sponsors the Hong Kong Productivity Council, the Vocational Training Council, and, since 1993, the Industrial Technology Centre designed to provide floor space for young technology-based companies, but plans for a science park have yet to be realized.

These arms-length approaches, the sponsoring of councils which aim to become self-financing, is the closest Hong Kong comes to state intervention. In a small economy such as Hong Kong, the role of government as a supplier and user of information could be a significant catalyst in the spread of information technology and services, but as of the beginning of 1995 only two government databases, covering law and land registration, were online to the private sector. The remit of the government's own Information Technology Services Department (ITSD), for example, is confined to improving government efficiency through the use of office automation, and not with the promotion of online services to the public. (See Greenfield and Lee, 1992.) Of all the initiatives outlined, the EDI scheme, when it eventually gets off the ground, is likely to have the greatest single impact in diffusing the domestic business use of online communications. The struggle of the new telecommunications entrants to establish a value-added services market will also be a driving force.

Getting things wrong is perhaps one of the lessons for the government in telecommunications technologies. From the late 1980s the government's Recreation and Culture Branch was determined to establish a cable TV network as a means of creating a second telecommunications network. Not only did this policy conflate technological synergy with business synergy[30] but it also embraced the British model of duopoly competition just as Britain was abandoning it. (See Chapter 6 for a further discussion.) The scheme collapsed when the consortium awarded the licence, which included the American Baby Bell company US West, split up in disagreement and disarray. Sensibly, the government then abandoned the duopoly model for Hong Kong by separating the cable TV licence from new entry into domestic telecommunications after 1995. If the cable system becomes a commercially successful territory-wide network it will then be available as a residential market for telecommunications services, interconnecting with operating companies. In a parallel development, Hongkong Telecom is testing a Video-On-Demand (VOD) system in preparation for the forthcoming age of competition in the multi-media market.

Besides multi-media, the new entrants are likely to target the business market for higher value-added sales and for international traffic. However, the most important source of international revenues comes from Hong Kong-China traffic which accounts for nearly 50 per cent of Hong Kong's outbound calls (with 80 per cent going to Guangdong Province), and over 30 per cent of international revenues. It is also the fastest growing traffic stream, limited only by the capacity of China's own network and the state of China's economy. According to an agreement of the Joint Liaison Committee, which oversees the arrangements of Hong Kong's return to Chinese sovereignty in 1997, Hong Kong-China traffic will still be treated as if it were international, an arrangement that suits China's Ministry of Posts and Telecommunications, the MPT, because it maintains the accounting rate regime which applies to international traffic. Usually the accounting rate, which fixes the per-minute payment a sending administration pays the receiving administration, is split 50:50, known as the settlement rate. The settlement rate on traffic to and from Shenzhen, the special economic zone bordering Hong Kong, is indeed 50:50, but with the rest of China the rate is 33.3:66.7 in China's favour. These net dollar payments to China are an important source of finance for China's own network expansion, and this in turn benefits Hongkong Telecom by increasing the capacity of China to send and receive traffic. But it also keeps IDD tariffs well above costs.

The arrangement also suits the Hong Kong authorities in the sense that it appears to uphold the 'One Country, Two Systems' principle which guarantees Hong Kong's relative autonomy as a Special Administrative Region (SAR). But will it satisfy the Guangdong Posts and Telecommunications Authority who only receive a fraction of the MPT's revenue from Hong Kong? Will it satisfy the new entrants which want to compete with Hongkong Telecom for China traffic, but cannot do so until after 2006 if it is treated as international? Will it suit CITIC, China's overseas investment arm, which has a 'little under' 20 per cent shareholding in Hongkong Telecom? Sooner or later, after 1997, traffic between Hong Kong and mainland China will be treated for what it really is, long-distance trunk traffic. See Ure (1995c) for a discussion. Before, or after, 2006 it will be open to competition, just as China's domestic long-distance market is opening to competition. Then the big question becomes: will Hong Kong-based companies be regarded as Chinese for purposes of direct investment in networks and services in China after 1997? These are part of the post-1997 unknowns.

Macau

Macau returns to Chinese sovereignty in 1999, just two years later than Hong Kong. It is a tiny Portuguese colony of less than 500,000 inhabitants on the mouth of the Pearl River delta, one hour's ferry ride from Hong Kong, and economically totally dependent upon Hong Kong and China. (See Sit, Cremer and Wong, 1991). Its many small and medium-sized workshops stand at the end of a long sub-sub-contracting chain which stretches through Hong Kong to the world economy, although its single most important source of revenue is from gambling, with tourism and shopping from Hong Kong and China's Zhuhai Special Economic Zone also important. Plans to build a new freight airport and land reclamation projects are being promoted to strengthen Macau's entrepot role. In recent years around 60 per cent of its telecommunications traffic was with Hong Kong and 30 per cent with China.

Since 1981, Macau's telecommunications services have been run on a twenty-year Build-Operate-Transfer (BOT) franchise by the CTM, Companhia de Telecommunicacoes de Macau, which is majority owned by Cable and Wireless plc., with CPRM of Portugal and China's CITIC as minority partners. In 1981, Macau had a teledensity of just 5 per cent which ten years later had jumped to over 22 per cent.

Singapore

Telecommunications in Singapore is very much part of the government's strategic vision for an information-telecoms industry which will turn the city-state into an Intelligent Island by 2000 (IT2000).[31] *The Next Step,* published in 1991, laid down the vision of Singapore as a fully developed nation within twenty to thirty years. Accordingly the Ministry of Trade and Industry published a *Strategic Economic Plan* charting Singapore's next decade of growth as a 'total business hub' with an emphasis on high value-added, knowledge-based industries and the internationalization of local firms.

Since separation from the Malaysian Federation in 1965, Singapore has been dominated by the Progressive People's Party (PPP) and the de facto head of state, Senior Minister Lee Kwan Yew. The PPP has led a proactive interventionist government which very consciously tries to steer the economy, and the society, towards industrial goals and productivity targets. Not surprisingly, therefore, the Telecommunications Authority of Singapore (TAS), which acts as the regulator, and the state-owned (and since 1993 partially privatized) monopoly PSTN operator, Singapore Telecom (ST), are both instruments of government industrial policy. For example, the partial privatization of Singapore Telecom in 1993 was in large part designed to strengthen the capital base of the Singapore Stock Exchange, see Hukill (1994).[32]

This policy includes local high-technology initiatives and overseas development projects pioneered by Singapore Telecom International (STI) to promote outward investment. Local projects include an island-wide ISDN network, begun in 1989, the National Computer Board's National Information Infrastructure (part of IT2000) for island-wide computer networking, the Economic Development Board's (EDB) promotion of the telecommunications industry as a leading industrial sector, and the 1991 National Technology Plan of the National Science and Technology Board (NSTB) which provides US$1.2 billion for science and technology parks, with ST responsible for installing broadband networks capable of simultaneous transmissions of voice, data and video.

ST will also install an optical fibre local loop for Singapore's cable TV network linking the island's 600,000 homes, while the Singapore Broadcasting Corporation (SBC) will install the inside cabling. Under a plan brokered by the Prime Minister's Office and the EDB, the cable TV network will replace the SBC's unsuccessful MMDS (Microwave Multipoint Distribution System) network known as Singapore Cable Vision (SCV), a consortium involving the Singapore Housing Development Board which had separate plans to cable its 7,000 public housing blocks. The plan forces the SBC to share facilities with ST and abandon its separate plans, but in typical Singaporean fashion the underlying tensions are submerged beneath the rationale of the state plan. With its plans for Video-On-Demand, ST may well emerge as a competitor of SBC. ST is also involved through a joint venture company, Singapore Network Services (SNS), in promoting a videotex service called Teleview, E-mail services including numerous information links, and a series of trading networks as part of the state-sponsored EDI project, TradeNet. Singnet was introduced in 1994 giving dial-up access to Internet over leased circuits.

Internationally STI is installing, in partnership with PT Telkom, the telecommunications infrastructure on Batam Island as part of the golden-triangle concept linking southern Malaysia, Singapore and the Riau Province of Indonesia. STI also operates Indonesia's first foreign venture cellular mobile telephone service, providing GSM coverage across Batam Island from radio transmitters in Singapore, although PT Telkom has moved smartly

Table 2.3
Singapore Telecom's Overseas Subsidiary and Associated Companies, 1994

Country	Company	Service
Australia	Infolink Network Services	Online services
China	JV with Beijing PTT	Paging
Hong Kong/China	APT Satellite Company	Satellite
Hong Kong	Sky Telecom Services Ltd	Paging
Indonesia	Batam and Bintang Islands	GSM cellular mobile
Indonesia	PT SkyTelindo Services	Skyphone
Malaysia	Information Network Services	Data comms and VAS
Malaysia	Intergated Information Bnd	Directory space
Malaysia	Multi-Media Communications	Multi-media equipment
Malaysia	Sudong Sdn.	Terminal equip.
Malaysia	Tourism Publication Corp.	Publishing
Mauritius	Teleservices Ltd	Paging
Norway	Netcom GSM A/S	GSM cellular mobile
Pakistan	Singapark Communications	GSM cellular mobile
Philippines	Globe Mackay Cable & Radio	Telecoms services
Sri Lanka	Lanka Cellular Services	GSM cellular mobile
Sri Lanka	Lanka Communications Services	Data comms.
Sri Lanka	MTV Channel	Broadcast services
Thailand	Shinawatra Datacom Ltd	Data comms.
Thailand	Shinawatra Paging Co. Ltd	Paging
Thailand	Infolink Co. Ltd	Paging
Thailand	Phonelink Co. Ltd	Paging
UK	Cambridge Cable Ltd	Cable TV/Telecom
UK	Anglia Cable Ltd	Cable TV/Telecom
UK	Yorkshire Cable Ltd	Cable TV
Vietnam	Saigon Mobile Phone Centre	GSM cellular mobile
Vietnam	Radio Paging Centre/Phonelink	Paging

to build their own GSM base station. Table 2.3 summarizes STI overseas ventures up to mid-1994.

'As of March 1993, the Group invested S$278 million (US$174 million) in 23 joint ventures and strategic investments in 11 countries in the ASEAN region, Asia Pacific, Europe and the United States' (STI Annual Report, 1992/3, p.47). By mid-1994, this had risen to S$627 million (US$392 million) according to *The Economist*, 4 June 1994, p.68. The same month STI announced the formation of Singapore Telecom Europe (ST Europe) to promote Singapore as a regional telecommunications hub. See *Telenews Asia*, 14 July 1994, p.14.

The opening of the Asia-Pacific Cable (APC) in 1993 for the first time provided Singapore with optical fibre links across the Pacific, and the opening of the ASEAN cable in 1994 will complete Singapore's optical fibre connections to all its neighbours, thereby catching up with Hong Kong. According to the Annual Report 1992/3 over 3,000 international companies hub through Singapore, but fewer are regionally based in Singapore than in Hong Kong.

Every utility service company and major industrial corporation in Singapore is included in, or is expected to adhere to, the government's strategic vision for the island. This includes a stress upon overseas investment to strengthen Singapore's regional role. For this reason Singapore places considerable emphasis upon its role as a regional telecommunications hub, which inevitably means competition with Hong Kong. (See also Chapter 6.) For example, ST's IDD rates are determined not by consideration of local cross-subsidy or cost-based pricing, but are set directly in line with Hong Kong's. To gain a competitive advantage in the high-quality and high-speed international data market, essential to multinational corporations, STI was a founding member of the AT&T and KDD WorldPartners global network management consortium, a move that persuaded Hongkong Telecom to follow suit despite being part of the rival Cable & Wireless Global Highway. (See endnote 26.) A further example of this strategy is Singapore's plan to launch its own satellites by 1999, with the possible involvement of Taiwan. Singapore is proposing to its ASEAN partners and to China that through these satellites an Asian regional mobile communications network be established, hubbing through Singapore. A learn Mandarin campaign at home is also being promoted to build bridges with China, where the Singapore government has been actively pursuing commercial projects, including telecommunications.[33]

Despite its internationalism, Singapore's telecommunications sector has been mainly closed to foreign direct investment until now. Following the 1992 Telecommunications Act which incorporated ST, the TAS granted ST an exclusive 15-year franchise on the PSTN, and a 5-year exclusivity on cellular mobile. ST introduced an analogue AMPS 800 MHz system in 1988, an ETACS 900 MHz system in 1991, and started a digital GSM network in 1994, acquiring 20,000 users by mid-year. But also in 1994, TAS confirmed that a second GSM licence would be awarded, operational from 1997 for 20 years, and the bidding includes overseas partners.[34] Then no new licence will be issued for a further three years. TAS also opened international bidding for three additional 10-year paging licenses. ST introduced its CT2 Callzone service in 1993, and in addition to supporting over 650,000 pagers, ST also offers Skypager (with Mtel or Mobile Telecoms Technologies Corp) and Skyphone (with BT and Norwegian Telecom) services for use during air travel. In preparation for greater competition, initially in the value-added services market, ST has formed Mobilelink to run its cellular business and Pagelink to run its CT2, paging and mobile data service, Dataroam which is offered through a joint venture, ST Mobile Data, with the US company Bell South. A further area of liberalization is to permit private owners of multi-tenanted buildings to purchase and install their own PABX and rent services to all tenants. Previously, each tenant had to buy or rent a separate PABX from ST.

Private operators in Singapore are limited to value-added services. Lines Technology, a subsidiary of the French company Matra, offers videotex through PCs to data-bases in France. Other dial-up PC data network international gateways, for example CompuServe, are available which prove commercially more attractive than SNS's Teleview information services. Other entrants include NETS (Network for Electronic Transfers Singapore) which was established by four local banks to provide ETF (electronic transfer of funds) for banks and POS (point-of-sale) services for shops. SNS, in which ST holds a 15 per cent share, offers, alongside EDI, a series of Nets and Links, such as MediNet, AutoNet, MailLink (E-mail) and OrderLink (for procurement).

In a city-state of under 3 million people, it could be argued that one public network is efficient and sufficient, but ST's reputation for being bureaucratic with customers, according to Tim Cureton, Hongkong Bank Group Telecommunication Manager (see Cureton, 1992) is one drawback. And other members of the General Agreement on Trade-in-Services (GATS) and the World Trade Organization may feel Singapore is using double standards when it is so actively attempting entry into other Asian markets.

South Korea

South Korea has the eighth largest telecommunications network in the world. In 1993 an exchange capacity of 20 million mainlines served 16.6 million subscribers from a population of around 45 million, giving a teledensity of nearly 38. Yet in 1981 only 8.4 per cent of the population had a telephone, and exchange capacity was under 3.5 million.

Until the 1980s, telecommunications in South Korea was a monopoly run by the Korean Telecommunications Authority (KTA) as part of the Ministry of Information and Communications (MIC, previously the Ministry of Communications). Under the pressure of a geo-politically divided Korea, governments in the South since the 1960s have relied heavily upon a strategy of export-promoted economic growth using the state to mobilize capital resources, and, as Park (1992, p.71) says, 'the economic development plan driven by the Government from the beginning of the 1960s resulted in the fast construction of the infrastructure of Korean telecommunications.' (See also Choo and Kang, 1994.) But not fast enough. By the early 1980s four key problems convinced the government of the necessity of telecommunications reform: a waiting list of half-a-million for the basic telephone service, successful calling rates often well below 50 per cent, outdated rural exchanges, and a lack of modern data communications. At issue was the competitive advantage of Korea's national economy.

The government's first steps came with the Interim Law for Expansion of Public Telecom Facilities in 1979 which increased tariffs and introduced subscriber bonds to raise finance for network expansion. Loans from the USA, Belgium and Sweden were also secured, and, as Choi (1993) records, 'with the start of the 5th Five Year Economic Plan (1982–87) the government gave investment priority to telecommunications infrastructure.' Investment in telecoms as a percentage of the government's overall fixed investment rose from under 3 per cent in the 1970s to 5.7 per cent in the 1980s. In parallel to network expansion the government also embarked on a strategy of building a research and development capacity to stimulate the local design, manufacture and export of switching, transmission and peripheral equipment. Funding responsibility was given to the KTA, while responsibility for the R&D itself was given to ETRI (Electronics and Telecommunications Research Institute).[35]

The first steps to liberalize CPE came in 1980 when subscribers were permitted to freely purchase their own telephone sets. In 1982 the KTA was separated from the Ministry and established as the monopoly provider of PSTN services, together with a subsidiary DACOM (Data Communications Corporation) as provider of computer-based data services. Another subsidiary, the Korean Mobile Telecommunications Corporation (KMTC), began operations in 1988 to provide mobile cellular (800MHz AMPS) and paging services. In the same year the Korean Port Telephone Company (KTP) was set up to provide seaport and airport communications. These moves were paralleled by an opening of the VANS market, beginning with database and data-processing services in 1985, and a full opening in 1989, so 'by the end of the 1980s, there were 171 telecoms service providers, including 5 common carriers such as KTA, DACOM and KMTC' (Hahn, 1992). For a useful review of South Korea's legislation and reforms see APT (1994).

Increasing US trade pressure on Korea was one reason for reform, leading to a bilateral agreement for a partial opening of markets.[36] South Korea launched into an opening of its network and service markets with legislation in 1989 and 1991 amending the 1983 Telecommunications Basic Law and Public Telecommunications Business Law. In preparation for competition a new regulatory body, the Korea Communications Commission (KCC) was established, but owing to insufficient independence and resources has failed to operate, in sharp contrast to Hong Kong. The KTA was incorporated as Korea Telecom (KT) as a step towards privatization. This began in 1993, and KT was required to divest its shareholding in DACOM which became South Korea's second international telecoms carrier in 1990. DACOM began its operations in December 1991.

DACOM was given encouragement by being allowed to offer IDD at 5 per cent discount on KT's tariffs, and by-pass on the domestic link, connecting the customer directly to the DACOM gateway by means of a circuit leased from KT.[37] At the same time, KT was allowed to enter the data communications business through its subsidiary Korea PC Communications. From January 1994 the distinction between domestic leased voice and data circuits was abolished. This permits KT to offer leased data circuits directly to the public without going through DACOM. Despite DACOM, South Korea's IDD rates remain high to protect domestic cross-subsidy and the stock market valuation of KT. The government plans to allow stock trading from 1995, and to sell off its own holdings in tranches of 10 per cent until the public own 49 per cent of KT. Market entry into the domestic long-distance trunk market is the next policy issue to be debated, but KT argues for tariff rebalancing to bring down IDD and trunk call charges and raise local charges, before competitors are allowed in.

The reforms introduced three new service categories: Network Service Providers (NSPs) which were divided into General Service Providers (GSPs) and Specific Service Providers (SSPs), and thirdly, Value-Added Service providers (VSPs). GSPs, namely KTA and DACOM, were designated as operators allowed to offer nationwide network services, with associated responsibilities for equipment and systems research and development. By 1993, the KTA was operating seven research laboratories. SSPs, such as KMTC and Korean Port Telephone (KPT), require approval from the MIC and may only offer specialized services, such as mobile cellular, paging and port telephone. The MIC seems to favour a duopoly model of network competition. In the paging market, for example, the KMTC provides a national service and, since the beginning of 1993, in each of the nine regions (including the capital city Seoul, and Pusan, the second largest seaport city) one other operator is allowed to compete, and in Seoul an additional operator, making for three in Seoul and eleven in total. According to the Korea Annual (1994) there were 2.6 million pagers in use at the beginning of 1994, 85 per cent of them with KMTC.

After an earlier failure,[38] the KMTC or KMT was privatized in 1994, with control passing to the Sunkyong Group, Korea's fifth largest business conglomerate or 'chaebol'. KMT has since become a signatory to the Motorola-sponsored Iridium project, while DACOM has linked up with the rival Globalstar project.[39] In April 1994, a second cellular licence was awarded to Sinsegi Mobile, a subsidiary of the giant Pohang Iron and Steel Corporation, after the government requested a recommendation from the Federation of Korean Industries. Foreign shareholding is limited to 20 per cent, although to assuage US demands an additional 2 per cent was offered.[40] The Korea Annual (1994) gives the number of cellular telephones by the end of 1993 as 470,000, a growth of 74 per cent in one year. Eighty per cent are handheld and 20 per cent portable carphones. Korea's cellular standard is currently AMPS, but ETRI has announced that CDMA will be adopted as a digital standard for the future, which opens the prospects for several digital standards to be adopted by Asian countries.

While foreign direct investment in NSPs remains restricted to 33.3 per cent in the case of any SSP, it has been lifted for VSPs which provide services such as E-mail, EDI, packet-switching, database information services, and so on. However, the rigid differentiation between GSPs and SSPs is being criticized in Korea as discouraging service innovation, and hampering operator economies of scale, while limitations on foreign investment is subject to bi-lateral Korea-US trade talks. At the same time the more liberal policy towards VANS is balanced by strong state involvement in the informatization of Korea's national economy and society. In this regard Korea is similar to both Singapore and Taiwan, although liberalizing more rapidly.

To foster a socio-scientific approach to the encouragement of an information economy and society, the government in 1985 established the Institute for Communications Research (from the policy arm of the Korea Research Institute of Telecommunications, established 1970) which, by Act of the National Assembly in 1989, changed its name to the Korea Information Society Development Institute, or KISDI. With nearly 70 research staff KISDI is a major think-tank providing input to the seventh Economic Social Development Five Year Plan and the indicative long-term investment and development plan of the ministry. KISDI is also venturing overseas in its contract work, and so is Korea Telecom in its investments, showing signs of emulating Hongkong Telecom and Singapore Telecom.[41]

The ministry has also established a 'Comprehensive Measure for the Information Society' (see APEC, 1993) and in this regard an equally important body in the informatization of Korea is the National Computerization Agency. The NCA has responsibility, under the Information Communications Bureau (ICB) of the ministry, for the promotion of National Basic Information Systems or NBIS which began in 1987. This is a networking project similar to Singapore's National Information Infrastructure under the IT2000 project. The Korea Research Environment Open Network (KREONNet) is part of NBIS, designed as a backbone to connect local computer networks in Korea's many research institutes. The Government Administrative Information System (GAIS), which predates NBIS but has become part of it, provides online databases offering information on topics such as housing, land, employment, vehicle registration, customs clearance, and so on. To stimulate usage the ministry's aim, according to Hahn (1992), the Director of the ICB's Planning Division, is to realize 'one PC per household' by 2000, 3 million supplied by the public sector and 7 million by the private market.

Other NBIS networks include the Financial and Banking Information Networking System, and the Korean Trade Association's EDI network, Korea Trade Network (KTNet). And, again in common with Singapore and Taiwan, the Korean government has tried to promote a nationwide videotex service. From 1985, DACOM launched Cheol-lee-Ahn ('Clairvoyant'), but through its subsidiary, Korea PC Communications, KT now provides the leading service, called HiTel. However, experience is showing that videotex, at least in this top-down utility form, is really a non-starter compared with alternative PC-based information systems, see Kuo and Ho (1995). And, not to be outdone, Korea has its own version of the Information Superhighway. Since 1993 the MIC proposed to build Korea's own broadband fibre-optic network, utilizing the subscriber portion of KT's national cabling plans. An association is being formed of telecoms operators, the NCB, ETRI and others from the equipment industry, and from the universities and research institutes. The information superhighway is to be financed by further sales of KT stocks.

The Korean strategy can really be summed up as:
1. promoting the informatization of industry,
2. the industrialization of information,
3. the liberalization of telecommunications as the platform upon which computer-based information networks and systems will interconnect and operate.

To this end Korea is pushing forward with plans to build nationwide competing public switched data networks, intelligent networks and ISDN facilities.

The plan envisaged by KT in the 1980s was:
- 1984–91: the digitalization of the network (93 per cent in 1992)
- 1992–96: introduction of commercial ISDN
- after 1997: ISDN becoming nationally available alongside national information networks.

Optical fibre cabling is planned as follows:
- 1992–96: up to 800,000 users
- 1997–2000: fibre into high-rise buildings
- after 2006: fibre to the home (FTTH)

In 1995 South Korea plans to launch its own satellite, Koreasat, and Vsat system to provide the infrastructure for media and data communications to complement its ground fibre and ISDN network facilities. See Son et al. (1991).

Taiwan

Since the Kuomintang (KMT) or 'Nationalist Party' retreated from mainland China to Taiwan in 1949 the island's security and vital communication links have been firmly under the control of the state. But Taiwan's transformation as one of Asia's four little dragons is bringing profound changes in economic, political and social attitudes. Trade, investment and communication with mainland China are being liberalized. Although direct contacts with the mainland are still restricted, following the lifting of martial law telephone calls to China, routed through the United States, Hong Kong and Japan, have been allowed since 1989, and they already account for more than 14 per cent of Taiwan's international traffic.

Under the 1950 Telecommunications Act, the Ministry of Transport and Communications (MOTC) issues network and service licenses, while the public switched telephone network is operated and regulated by the Directorate-General of Telecommunications (DGT). As in Korea with Korea Telecom, the DGT also acts as an arm of government industrial policy, for example, by promoting local equipment manufacture. See Shih and Lin (1994).

A study of telecommunications law in 1986 by the Council for Economic Planning and Development (CEPD) of the Executive Yuan (or 'Council') initiated moves to liberalize the market. Customer premises equipment (CPE) markets were opened to competition in 1988, and since 1990 over thirty private companies have been licensed to provide the following value-added services: EDI,[42] E-mail, voice-mail, accessing remote data-bases, data retrieval, and remote transactions. However, all circuits must be leased from the DGT, resale is prohibited, as is the mixing of data and voice over one circuit, known as data-over-voice. A private Value-Added Network Association is lobbying to widen the scope of competition to include items such as fax and packet-switched data.

In 1977, a proposal to split the regulatory function of the DGT from its operational role was defeated, and since then the DGT has successfully resisted challenges to its monopoly.[43] The profitability of telecommunications to the coffers of KMT governments has been substantial. In 1993 the DGT earned a surplus of NT$31 billion, of which NT$25 billion was taken by the Treasury. The rate of growth of telephone main lines has been modest during the past decade, averaging less than 10 per cent a year, leading to a teledensity of 37 lines per 100 inhabitants by 1993, for a population of around 21 million. This is slightly ahead of South Korea but far behind Hong Kong and Singapore.

Over the past forty years, due to Taiwan's precarious political and military security, KMT governments were reluctant to spend heavily on infrastructure; instead they run up the world's largest reserves of gold and foreign currencies. But now, like the other dragons, Taiwan has reached a milestone in its development as the government looks towards high-technology and high-valued added industries to sustain its economic miracle. The importance of telecommunications to Taiwan's rapid economic growth during the 1970s was recognized in the DGT's six-year Telecommunication Modernization Plan implemented in 1976, which laid special emphasis upon long-distance and international communications, and again in the national ten-year Long-Range Economic Planning Program, 1980–89, which incorporated the DGT's Ten-Year Telecommunications Development Plan. Installation fees were slashed by nearly 60 per cent to bring tariffs, if not supply, within the reach of the majority of people. At the same time Taiwan listed the information industry sector as strategic to the island's economic security, blaming its lagging performance on the poor state of telecommunications infrastructure.[44]

Telecommunications was one of the government's fourteen major construction priorities in 1985, following the Telecommunications Modernization Plan approved by the Executive Yuan in 1984 which targeted ISDN development and a teledensity of 50 mainlines per 100 inhabitants by the year 2000. More recently, infrastructure expenditure, including telecommunications, has been given a high profile in the country's Six-Year National Development Plan (1991–96), which aims to raise GDP per capita income to US$14,000 (CEPD, 1992).

But paper plans are not quite real. What happens is the DGT plans on a four-year basis and reinvents the figures for the Six-Year Plans, and DGT investments ebb and flow according to central government finances. The tenth Four-Year Plan (1990–93), which cost US$8.6 billion, completed a Phone-in-Every-Village programme. The eleventh Four-Year Plan, as well as the Six-Year Plan, calls for nine special projects, including 94 per cent local loop and 100 per cent international digitalization, a trunk network of national highway and submarine optical fibre cable, an Intelligent Network by 1994 and an experimental ISDN network by 1996 as a step towards broadband services by 2000 (DGT, 1993). The strategy, or vision, sees Taiwan becoming a communications hub for the region, serving sectors such as banking, shipping and tourism. But by late 1993 the US$4 billion earmarked for these nine projects, some of which had been completed under the tenth Four-Year Plan which overlaps with the Six-Year Plan, had been cut to US$3.3 billion due to government financial restrictions. Central planning in Taiwan, as perhaps in most countries, is more indicative than literal, but in Taiwan's case it hinders the pace of telecommunications development because resources are diverted.

Following the CEPD's initiative in 1986, the MOTC established a Steering Committee and three Working Groups to revise the Telecommunications Act. The result was three bills which have been attempting to wind their way through the labyrinth of the Executive Yuan and Legislative Yuan, only to be withdrawn in 1993 pending further revisions. Since the formation of an opposition party, the Democratic Progressive Party (DPP) in 1986, the KMT's monopoly on authority has been fractured and called into question. In addition, pressure from lobby groups and trade issues has been growing.[45]

The first bill proposed to follow the Japanese model (see Bruce, et al., 1988). It divides services into Type 1, which includes the network and basic services, and Type 2, which are value-added services dependent upon the facilities of the Type 1 carrier. The bill would also permit foreign direct investment for the first time into Taiwan's telecommunications industry. Foreign companies would be allowed to hold up to one-third share of equity in local joint-ventures. In a compromise with the Value-Added Network Association, a lobby group of private VAS operators, the MOTC has agreed that, once legislation is passed, the service categories would be reviewed every six months, implying progressive liberalization.[46] A further source of pressure has been the desire of Taiwan to join the GATT/WTO.[47] For example, while cellular mobile telephones remain classified as a Type 1 services, and hence the monopoly of the DGT, CT2 services have been opened to both domestic and, from December 1994, international operators, along with packet switched data services, store-and-forward facsimile, E-mail and international leased circuits for data and other value-added services.

The other bills would restructure the DGT as regulator, and hive off the operations of the PSTN to a new company, the Chunghwa Telecommunications Company (CTC), which is then to undergo partial privatization. The DGT is to remain an arm of government industrial policy, but its research and development functions would pass to CTC which will also take on an industrial policy role as part of the Government's three C's science and technology strategy: communications, computers, consumer electronics. Taiwan's science and technology policy mostly falls under the National Science Council, and every ministry has an Office of Science and Technology advisors, with the Science and Technology Advisory Group advising the Executive Yuan. The industrial, trade and economic development of technology, including information technology, is handled through the Ministry of Economic Affairs, and the Council for Economic Planning and Development of the Executive Yuan. Policy implementation is conducted through numerous government sponsored organizations such as the Industrial Development Bureau, the Science and Technology Information Centre, the Precision Instrument Development Centre, the Hsinchu Science-Based Industrial Park, and a series of non-governmental research institutes headed by the Industrial Technology Research Institute, or ITRI.

The DGT's research and development activities, as well as its procurement policies, are an integral part of this industrial strategy, and reforms to liberalize the sector will either compromise this role or will be compromised by it. The government in Taiwan is acutely aware of the competition its telecommunications equipment and components manufacturing faces from mainland China's expanding industrial base, in part financed by Taiwanese capital.[48] At the same time, concern with China is accelerating Taiwan's efforts to open further to the world economy, to internationalize, and telecommunications reform is at the top of this agenda.[49]

---

[1] An underlying hypothesis running through this book is that the underdevelopment of China — that is its lack of developed commercial law, regulatory enforcement and information markets — imposes major transactions costs which give smaller Sino-Asian companies temporary advantages in trade and investment over international companies. The contrast in the importance to Singapore and South Korea of telephone traffic to mainland China underscores the point that the lure of China has more to do with economic relations and proximity than with ethnic ties as such. But insofar as ethnic ties reduce transactions costs, then Chinese family and business links can offer a competitive advantage.

[2] China is said to have two rice bowls: one is iron, and is secure but holds little money, the other is clay which is easily shattered but holds lots of money. Families spread their gains and risk with one member holding onto a job in the state sector which gives housing, medical and other welfare benefits, while another goes for higher wages in the uncertain non-state sector. This is the new 'one family, two systems' model of urban family life. See *Inside China Mainland* (Taipei: Institute of Current China Studies) September 1993, v.15.9 issue 177, p.75.

[3] The Party has therefore set about creating the beginnings of a surrogate civil society through non-state organizations, such as housing associations, marketing organizations, professional societies, which Howell (1994a, 1994b) argues, remain dependent upon the state for finance, expertise, information, or for central or local government cooperation.

[4] Instructions issued by the Ministry of Public Security in early 1994 that require international networking to be registered, ostensibly to protect China's commercial and political secrets, reminds us that monitoring of international communications remains ubiquitous in the PRC. (See *Far Eastern Economic Review*, 10 March 1994, p.15). But controls are not necessarily political. The traditions of bureaucracy and employment heirarchy are also impediments to the notion of distributed access and usage of such tools of business as facsimile machines and personal computers.

[5] The situation is made worse by the many different types of non-automatic and automatic electro-mechanical and semi-electronic switches still is use. Under 40 per cent of cities of county level and above had automatic switching in 1980, although this had improved to over 85 per cent by the end of 1992.

[6] Di and Liu (in Mueller, 1994) report Chinese estimates of the economic internal rate of return on telecommunications investment in the PRC as 45 per cent over the 1980s.

[7] Jitong's corporate brochure for 1994 lists 25 shareholders, joint venture agreements with Pacific-Link (Hong Kong), Excellent (Malaysia) and Bell South (USA)

and a business contract with IBM which has resulted in the Jilong Information Network Research and Development Company. The Chairman of Liantong, Zhao Weichen, in an interview in *Telecommunications Development*, September 1994, pp.13–26, lists 13 shareholders and identifies Liantong's targets as raising the national telephone penetration rate by 1 per cent by 2000, to gain a 10 per cent share of national trunk traffic, and a 30 per cent share of the cellular mobile telephone market, and to achieve direct international connnection.

8   It is a tradition to term as 'golden' projects which are considered nationally strategic. They are numerous, the best known being the Golden Bridge information network, the Golden Card banking network, the Golden Gate customs network, and the Golden Cellular digital mobile telephone network. Others include the Golden Finance, Golden Tax, Golden Sea.

9   By late 1992, China's total fixed-wire network capacity was 32 million lines, of which, according to *China Business Weekly*, 21–27 February 1993, '19.26 million were for public use and 12.74 million for private or specialized use.' This suggests that non-PSTN networks, under the control of state bodies such as the Energy Resources Ministry, the Ministry of Railways, the PLA (People's Liberation Army), the state airlines, roads and port authorities, and so on, constitute 40 per cent of China's total.

10 The State Council has moved to tighten controls over the use of scarce, and therefore valuable, radio frequencies: 'the radio commissions of lesser cities will be superseded. The offices of the central, provincial, district or municipal radio commissions will be located in the MPT and the provincial or local PTAs. These will be independent institutions and not under the organization of the PTAs. The P&T departments of different levels are to observe strictly the National Regulations for the Control of Radio Communication under the direction of the Radio Commission.' Statement by MPT Minister Wu, *China Telecommunications Construction*, v.6.5, October 1994, p.14.

11 China reserves the 900 MHz waveband for public cellular mobile services, and the 800 MHz for non-public. The PLA controls a swathe of the 800 MHz frequencies, including the internationally used bandwidth for CT2 (864–868 MHz), which also overlaps with the trunk radio frequency in China (864–866 MHz). Apparently for security reasons, the State Radio Regulation Commission, against PLA wishes, rejected the adjacent waveband (866–870 MHz) and choose a non-standard frequency for CT2 (839–843 MHz) in China, thereby depriving the PLA of a business opportunity. Instead companies have to do business directly with local branches of the SRRC, which in turn may become a partner. An example of the socialist market economy in practice.

12 For a background to this rivalry see Zita (1987) and Xu (1994). Xu also provides a short history of the MPT, which lost its powers to local governments in 1958, regained them in 1961, was dissolved in favour of the Ministry of Communications under the control of the PLA in 1971, was reinstated in 1973 but with local governments leading the way, and in 1980 had its powers restored.

13 The conservatives were grouped around Premier Li Peng, and veteran Party economist and central planner Chen Yun. In this context, conservatism implied placing greater emphasis upon controlling inflation, less on double-digit economic growth, less on market reforms and more on plan-guided reforms. It was precisely this threat to market reform that goaded Deng into his Mao-like gesture of going to the country. Behind the thinking of both sides was the ability of the Party to maintain its control in the wake of the massacre in Beijing in 1989, and the subsequent threat to the Party's popular legitimacy. For a penetrating account of the political economy in China since 1978 see Skirk (1993).

14 For an account of science and technology policy in China, see Conroy (1992).

15 The State Commission for Science and Technology is primarily responsible for identifying national research targets, and the State Planning Commission for directing resources. Locally developed telecommunications technologies mainly involve the MPT, the MEI and the State Commission for Education. Chen Yunqian (IEEE *Communications Magazine*, July 1993, pp.20–22) lists the key universities and research institutes as the Beijing University of Posts and Telecommunications, Chengdu Electronic Science and Technology University, Shanghai Jiao-Tong University, Beijing Tsinghua University, Xi'an Electronic Science and Technology University, Xi'an Jiao-Tong University, Hangzhou Zhe-Jiang University, and the MPT's China Academy of Telecommunications Technologies and the Wuhan Research Institute of Posts and Telecommunications. The CATT in turn oversees fourteen MPT research institutes.

16 See Ding Lu in Mueller (1994) pp.195–205. The profits tax concession has since been removed, reverting back to 33.3 per cent, as part of overall state enterprise reform measures. See interview with MPT Minister Wu Jichuan, *China Telecommunications Construction*, v.6.5, October 1994, pp.6–18.

17 On the other hand, the PTAs carry the social overheads of the iron rice bowl, and it will take some years of state enterprise reform and the setting up of a social welfare system before those liabilities are off the books. One of the problems with this process as it affects the MPT is the valuation of assets such as workers' housing and dormitories, clinics and schools and the division of responsibilities for these between different levels of the MPT's national structure. The World Bank is funding joint consultancies with the MPT into these issues. They are also looking at tariff reform.

18 In cases of dispute, the Chinese version of the contract is upheld, which may be open to a very different interpretation from versions in English, French, German, Japanese, etc. Enforcement of adjudicated disputes is another problem. 'Guanxi' or 'who you know' is another problem.

19 In October 1994, Hongkong Telecom's chairman, Lord Young, announced joint plans with the MPT to construct a 3,000 km fibre cable from Beijing to Hong Kong and with the Beijing PTA to install a GSM system, hinting these deals implied more than a building contract. 'No one before has ever joined China in a fibre network or a mobile phone network. . . . We are into running systems

and we are not going to change that policy.' (*South China Morning Post*, 13 October 1994, p.B1). He stopped short of claiming an equity agreement.

20 'So far, China has built 12 major computer application systems, 100 information networks and 800 databanks which have stored 50 million items of data. The information industry achieves an annual sales volume of 8 billion yuan (US$930 million).' *China Daily*, 28 September 1994, p.11. But according to the *China Daily*, 2 October 1994 , p.1,'only 300 of 800 databanks are in operation, of which less than 60 really serve the public.' According to the Hong Kong *South China Morning Post*, 14 June 1994, the research company, Dataquest, estimates that still 90 per cent of computers are bought by government-related companies.

21 In 1988, the State Science and Technology Commission launched the 'Torch Programme' to accelerate the commercialization of technologies, including the use of computers. See 'China Computers' by Saiman Hui and Hilary B.McKown in *The China Business Review*, September-October 1993, pp.14–24.

22 The MPT still consolidates the posts and the telecommunications costs and revenues, but at least provides separate reference to postal revenues. In 1993, total operating revenues were 31.2 billion yuan telecoms and 8.03 billion yuan posts. *China Posts and Telecommunications*, pp.3 and 9, Beijing: MPT.

23 Comparison of GDP's for 1991 show Hong Kong at around US$80 billion and Singapore at around S$27 billion. Hong Kong as yet has no measure of GNP.

24 For an analysis of the role of Hong Kong's small and medium-sized enterprises, see Sit and Wong (1989).

25 For detailed accounts of Hong Kong's international and domestic telecommunications networks, their histories, technologies and surrounding policy issues, see Mueller (1993) and Ure (1992).

26 This may be seen as a strategy by C&W to reduce the risk of losing control of Hongkong Telecom at any time in the future, as well as offering flexibility in approaching potential Asian partners. Significantly, Hongkong Telecom eventually decided to join the AT&T-KDD led managed data networking alliance WorldPartners, which includes Singapore Telecom, despite other subsidiary and affiliated companies of C&W, such as Mercury in the UK, Optus in Australia and IDC in Japan rejecting the option. WorldPartners stands in rivalry to Concert, a consortium of BT and MCI, to the alliance between US Sprint, Deutsche Bundespost Telekom and France Telecom, and to Cable & Wireless' own 'Global Digital Highway'. Unisource, another European alliance between the Dutch, Swedish and Swiss PTTs, also decided to join WorldPartners in 1994. As part of the C&W/Hongkong Telecom strategy to expand regionally, the group is leveraging the competitive strength of C&W in cable laying, and Hongkong Telecom's experience in cellular mobile communications. (For more details, see Chapter 5, especially Table 5.8.)

27 This phrase was coined by a senior government official, Sir Philip Haddon-Cave (1989). For details of what this policy means in practice, see Chen and Li (1991).

28   The idea behind price-incentive regulation is that the company has an incentive to increase its efficiency by lowering its costs by more than X, such that when the weighted basket of tariffs is allowed to rise by the price index minus X, the company increases its profits margin while the 'average' consumer enjoys falling real prices. For a discussion by Littlechild who developed the concept, see Beesley and Littlechild (1989).

29   Estimates of computers in personal use come from Survey Research HongKong Ltd. Estimates of business computers come from Datapro and information received from Graham Mead Associates, Hong Kong.

30   Telecommunications and television are very different businesses, with very different investment profiles and producing very different services according to very different timeframes, to say nothing of very different management styles which are usually associated with these sectors. For some fine details of the saga, see Day (1994).

31   A National Information Technology Plan was formulated in 1985–86, involving the National Computer Board, the Institute of Systems Science, the Economic Development Board and Singapore Telecom. (See Kuo, 1994, and Wong, 1993.) Using input-output tables compiled by the Department of Statistics for 1983, Heng and Low (1990) found the information sector's contribution to income and output in Singapore 1973–83 was second only to the services sector, although by sucking in intermediate imports, such as computer and telecoms hardware and software, foreign exchange earnings were negative. They conclude that 'information is the powerhouse of future post-industrial growth, and this is one area that Singapore just simply cannot opt out of, especially with other NICs competing so closely in it.' (p.66). Earlier, Chen and Kuo (1985) found support for a 'reciprocal relationship between economic development and telephone availability and use' (p.243) in Singapore over the period 1964–82.

32   This had two purposes. It helps promote Singapore as a regional financial centre, and it spreads wealth-ownership, giving people 'something to fight for' as Lee Kwan Yew put it in an interview with the *South China Morning Post*, 27 November 1993. As a boost to Singapore Telecom's revenue prospects, and its potential share value, calling charges were introduced from January 1992. For details, see Kuo (1994).

33   Under the initiative of Senior Minister Lee Kwan Yew, who enjoys the status of a political advisor to the government in Beijing, Singapore has promoted the Suzhou Township Development Project. Suzhou is a town of less than one million people in the Yangste river basin, in Jiangsu province, close to Shanghai. Singapore is investing in a new town development, including the design and installation of a modern telecommunications network.

34   In a private interview TAS explained the overseas partner was to transfer experience in managing cellular projects to the Singaporean new entrant in preparation for promoting Singapore's capability to compete for similar projects within the region. A serious debate is taking place behind the scenes in Singapore

as to how far and how rapidly to shift from a protected to a competitive market model across the whole telecommunications sector.

35 For an account of South Korea's planning process, and especially the Fifth Five Year Plan which prioritized telecommunications and high technology development, see Song (1990). Also see Wade (1990) for South Korea's industrial policy towards the telecommunications sector. South Korea's science and technology policy derives from *The Long Term Science and Technology Development Plan Toward 2000*, released in 1986.

36 The Omnibus Trade Act, 1988, requires the US government to retaliate against countries refusing to abandon certain kinds of protectionism against US companies. From 1987 US trade representatives began Market Access Fact Finding (MAFF) talks with Korea over telecoms, leading to a more open market agreement in 1992. See Sung (1994), also Choi (1992) and Bruce and Cunard (1994).

37 Dacom's IDD tariff differential was later reduced to 3 per cent. As of the beginning of 1994 DACOM had won 30 per cent share of international traffic from Korea Telecom.

38 Following a public outcry the original award to Sunkyong was revoked because the son of Chey Jong Hyon, the Group's chairman, was the son-in-law of the then President Roh Tae-woo.

39 There are several proposals from US-based companies to launch multi-satellite mobile global communications systems. Motorola's Iridium project was the first to be announced. The main rivals seem to be Inmarsat-P; US Orbcomm's Odyssey project led by TRW with Teleglobe of Canada and including TRI of Malaysia; Spaceway led by Hughes, AT&T/ McCaw, Mtel and Singapore Telecom (ST is also proposing a regional satellite mobile communications network); and Globalstar, a joint venture led by the Loral Qualcom Corporation, and including the Korean electronics giant Hyundai. (For details of other projects see *TelecomAsia* v.6.2: February 1955: 24–27 and 48.) KMT claims exclusive coverage with Iridium over the whole of Korea, but clearly an accommodation with North Korea would be necessary.

40 Four US companies were offered shares: Airtouch Telecommunications Corporation (ATC, formerly Pactel), Southwestern Bell, GTE and Qualcom, but GTE withdrew in protest at being limited to 4 per cent. Foreign equity in SSPs is restricted to 33.3 per cent, but in this case only 20.2 per cent was available. A further 2.2 per cent given up by some of the original consortium shareholders and offered to GTE. GTE was not mollified.

41 In early 1994, KISDI secured a consultancy, with Asian Development Bank funding, to assist Indonesia's planning of an Integrated National Telecommunications Strategy Development project. This grew out of the twinning programme Korea Telecom adopted with Indonesia in the 1980s. See Yu Wan-Young and Pahng Sukbaum (1991). Also *Telenews Asia*, 28 July 1994, p.9, reports KT bought into Retelcom (Republic Telecoms Holdings) of the Philippines, which through the Santiago family, controls PT&T, Capwire,

Philippine Wireless and the paging company Pocketbell. And according to *Asian Communications*, October 1994, p.8, KT entered a fixed-wire network building contract in Vietnam. See also Chapter 3.

42 Taiwan distinguishes three models for EDI. The first is 100 per cent government ownership, such as Trade-Van for handling cargo manifests. The second is government sponsored systems, such as Trading Net for the wholesale and retail trades. The third is the autonomous network, but supported by the government, such as for the automobile sector. See Stewart Carter, editor of *The Australasian EDI Report*, in *TeleNews Asia* v.1.5, August 1993, p.11.

43 The corporatization of the DGT was subsequently reviewed in detail. See Tseng and Mao (1994) for a background. Tseng led a subsequent joint study by the Council for Economic Planning and Development (CEPD) and the China Interdisciplinary Association (CIDA) for sector liberalization: Tseng, F.T. et al. (1987). *Modernization of Telecommunications and Information: A Study of Related Regulations*. Unpublished, Taipei: CEPD.

44 See Wang, G. and F.T.Tseng (1992), The First Step to Privatizing Telecommunications Industry in Taiwan. Paper presented to the 9th International Conference of the International Telecommunications Society, 14–17 June: Sophia Antipolis, France. mimeo.

45 One source of pressure comes from the Taiwan Telecommunications Workers Union (TTWU) which Shih and Lin (1994, p.338) describe as 'primarily a benevolent association.' Seats in the National Assembly and the Legislative Yuan are reserved for the TTWU, although the union 'despite the technological fluency of its members, was not involved in the policy-making process.' In 1986, a TTWU member of the DGT ran as a DPP candidate 'and upset the KMT incumbent, who also happened to be the union's chairman . . . Another employee was elected DPP member of the National Assembly.' Shih and Lin conclude that although the DPP 'has not proclaimed any explicit objections or alternative proposals to current telecommunications policy, the underlying force of an opposition voice in the union cannot be neglected.' Fear of losing the benefits which go along with civil servant status, such as security of employment, pension rights and seniority rewards, are frequently a cause of friction when state departments are corporatized as state enterprises, and then partially or fully privatized.

46 See Chen, P.L. and C.C.Liu, G.Wang (1995), Liberalizing Telecommunication Services in Greater China: Competing, With Whom? Paper submitted to the Communication Law and Policy Interest Group, International Communications Association Conference, 25–29 May: Albuquerque, New Mexico, USA. mimeo.

47 As Bruce and Cunard (1994) say, 'Taiwan has been subject to less significant external and internal pressures than other countries to open the telecommunications sector.' (p.201), but the threat of increasing political isolation is shifting Taiwan's policy emphasis.

48 '[T]he *Economic Daily News* said the Mainland Affairs Council (MAC), the is-

land's top China policy-maker, has drafted measures to limit Taiwan firms' investments in mainland projects to 10 per cent of their paid-up capital . . . It said China had become Taiwan's main competitor in major markets abroad.' *South China Morning Post*, 22 November 1994 (Business Post, p.4)

[49] China's ongoing campaign to deny international recognition to Taiwan's political regime, known as the Republic of China, precludes Taiwan from direct membership of multilateral bodies such as Intelsat. Consequently, Taiwan has been searching for an Asian signatory of Intelsat to team up with. According to a report in the *Hong Kong Standard*, 29 October 1994, an agreement with Singapore seems to have been blocked, at least temporarily, by China's objections.

Telecommunications in ASEAN and Indochina

John Ure

> The foundation of ASEAN's success in the late 1970s and early 1980s were laid before 1975, in the era of 'containment', when the United States devoted all its energies to keeping Southeast Asia free from communism... One might argue further that the cohesion of ASEAN after 1975 owed as much to a combined commitment to face Vietnam, so as to pre-empt a communist irruption into Southeast Asia, as to any desire to fill the vacuum left by the United States.
> (Shibusawa, Ahmad and Bridges, 1989, p.87)

The raison d'être of ASEAN (Association of Southeast Asian Nations) turned more upon the aim of containing the spread of communism in Indochina than on an attempt to create a mutually supporting economic alliance of nations. With the end of the Cold War, Vietnam is now close to joining ASEAN, along with Laos, with Cambodia and Burma (Myanmar) waiting in the wings. ASEAN's efforts to define a new role for itself have mainly centred upon its conception of an AFTA (Asian Free Trade Area), but this creates tensions between Singapore, which already dominates the trade flows of the ASEAN countries,[1] and other members, notably Malaysia and Thailand, which are hesitant about exposing their economies.[2]

This underscores the point that ASEAN nations are at very different stages of their development, and finding areas of mutual economic and industrial interest is not easy. Rather than complementing one another, their economies compete in many ways for comparative advantage in terms of low wages, authoritarian controls over social movements, cheap land, and so on. What they share in common is often a highly bureaucratic approach to policy issues, and a strong sense of nationalism. With the exception of Singapore, they also share very low penetrations of telecommunications facilities. But like China and the dragon economies, all ASEAN countries now find the world market integral to their development prospects. Telecommunications is therefore a development issue for these countries, and this chapter will examine how the issue is being handle in each case.

Indonesia

The archipelago of the Republic of Indonesia sprawls 5,000 km across the equator, encompassing many different peoples and dialects, numerous religions including Islam, Hindu, Buddhism, Christian and pagan. It is the world's fifth most populous country, home to about 190 million people, with just under half of the 13,000 islands inhabited. The terrain ranges from lowland rice-paddy to remote jungle to mountainous, rugged and volcanic areas. By the end of 1992, the entire country was served by just 2 million telephone exchange lines, of which only 1.5 million were connected, around 0.8 mainlines per 100 inhabitants, one of the lowest teledensities of the Asia region. Forty-five per cent of lines are concentrated in the capital city, Jakarta, where less than 8 per cent of the people live, and the successful call ratio (SCR) is as low as 30 per cent for domestic dialing and 50 per cent for international dialing. And despite Indonesia's recent high rates of economic growth, telecommunications has played very little role.

At this stage in Indonesia's national development, the relationship between growth and telecommunications appears confined to selected areas of the economy and society. Kuntjoro-Jakti and Achjar (1993) demonstrate a correlation between telecommunications demand and the financial, tourism and trade sectors, but comparatively little from domestic resource-based manufacturing. At the same time, social demand is constrained by Indonesia's inequality of income distribution.[3] The World Bank's estimate of annual per capita income in 1991 was a lowly US$619. The rising middle and upper classes in Indonesia, say those receiving over US$1,000 a month, are probably less than 10 per cent of Jakarta's population where most of the nation's wealth is concentrated.[4]

National development and telecommunications have yet to be successfully linked, but the need is now recognized, leading to major reforms during 1994. The sector is subject to debate at the monthly advisory meetings of the National Telecommunications Council, chaired by the Minister of Tourism, Posts and Telecommunications (MTPT) and it figures in the national Repelita (Five-Year) Plans of the National Development and Planning Agency (BAPPENAS). However, in practice telecommunications objectives in the first five Repelita plans (1973–93), and the accompanying First Long-Term Development Plan, were little more than a restatement of the plans of the state-run domestic operator, PT Telkom (PT Telkomunikasi) and international operator, PT Indosat.

This may be about to change. In June 1994 the government officially announced, although apparently not without some dissent within the government,[5] a deregulation package, known popularly as PP20, which permits foreign direct investment in the industry up to 95 per cent, and overturns a previous constraint which required local companies to hold majority stakes. Behind this move is a plan to divide PT Telkom's operations into twelve geographical areas, with each operating company free to enter a Joint Operating Scheme (JOS) with private companies contracted to build turnkey local networks. In return for skills and technology transfer, the private JOS partners, which by implication will include foreign telecommunications companies, will share 70 per cent of installation fees and fixed monthly rentals, and 70 per cent of outgoing domestic and international call revenues generated from their networks.

As a parallel development in the controlled liberalization of the sector, PT Indosat was 32 per cent privatized in October 1994, raising US$1.1 billion of which 27 per cent will return to the industry, and PT Telkom announced similar intentions. So the commitment to development is showing signs of becoming a real process as a result of domestic and international imperatives. The domestic factors arise from the need to keep economic growth going to accommodate and balance different institutional — for example, military, religious — and social group interests which have been held together through a state ideology known as Pancasila, or the Five Principles. These are a belief in God, humanity, nationalism, representative government and social justice, principles which may be open to interpretation, but not to challenge. President Suharto's regime has not hesitated to use the authority of the state to enforce adherence to these concepts, regarded as vital to the New Order which was ushered into power after the bloody coup and counter-coup of 1965.

But with Indonesia's growing linkages to the world economy, and its position as a leading member of ASEAN, President Suharto has tried to balance the traditionalism of the state bureaucracy with a largely American-trained technocracy within the government. He has also carefully shifted the balance of representation among ministers and senior officials between the Christian minority and the Moslem majority. The government allows two official opposition parties to exist, but they are tightly controlled, to the point where the President approves their choice of leaders. ABRI, Indonesia's armed forces, remains the arbiter of political succession in Indonesia, but a real threat to social stability looms from the poverty of the vast majority of Indonesians and the inordinate wealth of the very rich. Around 70 per cent of the corporate wealth of the country is believed to be controlled by Sino-Indonesian families, although ethnic Chinese or Sino-Indonesians constitute only around 3 to 4 per cent of the population.[6]

When political society is closed to all but a small circle, a sharing of the privileges of power becomes necessary to avoid internecine rivalry. This has been achieved through control over large sectors of the country's resource-based wealth-producing industries by the government, by ABRI, and by the granting of monopolies to private conglomerates. Foreign companies wanting to invest in Indonesia or sell components, such as telecommunications equipment, also have to pass through the state machine, taking on local partners approved by the government and/or ABRI.[7]

As in China, the aim of national unity is, for the state, paramount. Telecommunications offers one means to this end, but economic nationalism is quite strident in Indonesia and at least until 1994 there was entrenched resistance to the opening of telecommunications to foreign investors. However, to make any substantial progress in building a national network there is a need to import equipment and the capital to support it. Local funds are insufficient and interest rates on loan capital have been prohibitive. Furthermore, the World Bank, which assisted in earlier projects, has been actively encouraging Indonesia to open its markets in return for supporting a Fourth Telecommunications Project. This includes funding for training Ministry of Tourism, Posts and Telecommunications (PARPOSTEL) officials, and the Directorate-General of Telecommunications (DGT) department within it, in the arts of financial and accounts management and in regulatory procedures in preparation for an opening to competition.

Under Indonesia's Five-Year Repelita Plans, which have a habit of overlapping to meet their targets, main lines have grown since 1971 at an annual rate of 11 per cent, and slightly higher since the 1980s. But Repelita 6 (1994–98) — which is Phase One of the Second Long-Term Development Plan (1994–2019) — aims to add a further 5 million lines, a CAGR of 26 per cent to bring teledensity to 3.2 per cent, and up to the ITU's average for a comparable per capita income. This increase of the targets reflects the government's determination to steer Indonesia into a newly industrializing country status and achieve a level of teledensity commensurate with the ITU's inter-country GNP per capita comparisons.

The use of more private capital to achieve this objective is enshrined in the Broad Outline of National Policy (Garis-garis Besar Haluan Negara or GBHN). President Suharto's stated objective is to achieve some level of equivalence with developed countries by the year 2019, an ambitious target compared with Malaysia's Vision 2020. Leading the charge has been the Ministry for Research and Development (BPPT), under the energetic leadership of B.Jusuf Habibie who champions Indonesia's leap-frogging into high tech industries such as aircraft manufacture, a policy which, for a country with a GNP per capita income of US$610, did not find favour in the World Bank's Annual Report.[8]

Indonesia has a major investment problem to achieve its telecommunications targets. PT Telkom has had the monopoly of domestic basic services, which is defined to include mobile voice communications, and until recently of Palapa, Indonesia's domestic and regional broadcast satellites, while PT Indosat enjoyed the monopoly of international communications. Both corporations are profitable, but the government takes around 60 per cent of pre-tax profits. So other sources of finance are required, and the 1994 changes in foreign investment law are designed to address this issue.

Indonesia's industrial laws fall into a hierarchy: laws by consent of the President, regulations issued by the government in the name of the President under ministerial signature, and decrees of Ministers. In 1989, Law Number 3 opened the door to the private sector. Nine local companies, some with foreign telecommunications companies as consultants, entered into PBH (Pola Basi Hasil) revenue-sharing agreements to build-out the network, transfer ownership to, and revenue-share with, PT Telkom. This arrangement may be regarded as a BT (Build-Transfer) as distinct from the BTO (Build-Transfer-Operate) arrangements in Thailand, and as such leaves little room for the private partner. See Table 3.1.

Table 3.1
PT Telkom's PBH (Pola Bagi Hasil) Partners
Revenue-Sharing BTs under Repelita V (1989-93)

PT Astra Graphia (with PT Telekomindo Prima Bhakti)
PT Bakrie Electronics (associated with Binmantara)
PT Dwi Mitra (associated with PT Telekomindo Prima Bhakti)
PT Elektrindo Nusantara (Binmantara Group)
PT Erakomindo Puranusa (Ciputra and Udinda Groups)
PT Propelat
PT Telekomindo Prima Bhakti (PT Telkom pension fund)
PT Wahana Esa Sembada
PT Wanaha Komunikatama

In 1993, Regulation Number 8 allowed private operation of value-added services, mostly data, on-line services, E-mail, but also paging. This was followed up by Decree Number 39 which for the first time raised the possibility of foreign companies entering the PBH scheme, and of joint ventures. These reforms were encouraged by the World Bank as a means of increasing both the rate of network build-out and the level of operational efficiency. The only genuine joint venture of 1993 was PT Satelindo, a company dominated by Bima Graha, a subsidiary of Binmantara which is closely associated with the President's second son, Bambang Trihatmodjo. PT Satelindo has been awarded control of Palapa, Indonesia's satellite network, and nationwide rights to operate a GSM cellular mobile telephone network. PT Ratelindo entered the other joint venture of 1993, which is really a revenue-sharing arrangement without equity ownership in the network. PT Ratelindo is connecting buildings in Jakarta and Bandung to PT Telkom's exchanges using a fixed cellular radio connection. Behind PT Ratelindo is the Bakrie Electronics company, again with connections to the Binmantara group.

During 1994, further moves towards forming joint ventures have begun in the cellular markets. Besides its pension fund company, PT Telekomindo, PT Telkom has entered revenue-sharing agreements with three analogue mobile operators (see Table 3.2) but these companies only share outgoing

Table 3.2
Telecommunications Companies in Indonesia 1994

Service	Company		
Domestic PTSN	PT Telkom		
	PT Ratelindo[1] to operate a radio local loop for PT Telkom in Jakarta on a revenue-sharing basis		
Payphones	PT Telkom		
	Over 1,000 Wartels (private kiosks and telecom offices) on a revenue-sharing basis with PT Telkom		
Intern'l Gateway	PT Indosat		
	PT Satelindo[2] (not yet operational)		
Cellular	PT Rajasa	470 MHz	Jakarta-Bandung
	PT Centralindo Panca Sarti	800 MHz	Bali, Java, Riau, Kilmantan, Sumatra, Sulawasi
	PT Elektrindo Nusantara[3]	800 MHz	Jakarta-Bandung
	PT Telekomindo	800 MHz	Kalimantan
Paging	23 private operators each limited to a maximum of 3 cities on a revenue-sharing basis with PT Telkom		
	PT Telematrixindo (nationwide) with Matrix		
	PT SkyTelindo (nationwide) with Singapore Telecom (also Skyphone)		
Data	PT Telkom PSDN		
	PT Citra Sari Makmur (with PT Telkom and Bell Atlantic) provides Vsat services		
	PT Lintasarta (with PT Telkom and PT Indosat) provides Vsat data-over-voice services to the banking sector.		
Satellite	PT Satelindo[2] licensed to operate the Palapa C generation of domestic and regional satellites. Palapa B also being handed over.		
Trunked Radio	PT Mobilkom Telekomindo in joint venture with Telstra		
	PT Maesa in joint venture with Cable & Wireless		

Notes: (1) PT Ratelindo is owned by Bakrie Electronics, with associations to Binmantara;
(2) PT Satelindo is a joint venture between PT Indosat (10%), PT Telkom (30%) and Binmantara (60%);
(3) PT Elektrindo Nusantara is part of the Binmantara Group.

call charges. They have had to depend upon the sale of handsets for their profits. Not surprisingly handset prices have been exorbitant. The arrangement with PT Telkom is a strange one. Rather than being for a fixed term of revenue-sharing it has been based upon a target NPV (net present value) of discounted future earnings, so the period of the agreements have varied according to interest rate changes and changes in the levels of capital outlay. Joint ventures based upon improved revenue-sharing arrangements are expected to bring handset prices down.

A model of regional duopolies has emerged for mobile telephony in Indonesia, with PT Telkom operating one network (known as STKB or Sambungan Telephone Kendaraan Bermotor — Mobile Telephone Connection System) and a joint venture including PT Telkom operating the second. The shift to digital mobile networks has begun with PT Telkom's operation of a GSM system on Batam Island and nearby Bintan Island. Ericsson equipment will be used in a joint venture with the country's telecommunications manufacturing arm, PT Inti. Duopoly is again the model, but nationwide this time, with the joint venture company PT Satelindo using Alcatel equipment in a joint venture with local partner PT Enkomindo on the second network.

Like other ASEAN countries, the Indonesian state has entered the Asia-Pacific era with high aspirations and a strong sense of nationalism. So until very recently there was in Indonesia's approach to foreign participation in the telecommunications sector a tension between the need to attract overseas investment and resistance to foreign ownership of what was still treated as a national sacred cow rather than as a basic necessity of modern economic and social life. The partial exception was the agreement with Singapore to develop the telecommunications infrastructure on Batam Island, just off the coast of Singapore, as part of the Batam-Singapore-Johor Bahru (Malaysia) development triangle set up to establish an offshore exporting capacity by attracting investment from Japanese, American and European multinationals. It may be more attractive for ASEAN countries to open their doors to each other first, and a number of ASEAN and local Asian companies are positioning themselves to bid for JOS contracts. The proliferation of areas designated as 'growth triangles' between ASEAN member states in particular may herald a political desire to enhance cooperation between Asian countries on a regional and sub-regional basis. The popularity among Asian governments of 'triangulation' is lending a new geometry to the economic geography. For example, an agreement between Indonesia, Malaysia and Thailand offers a tourist triangle, with the respective national airlines sharing part of the package. Tourism and hotels remain among the biggest generators of telecommunications traffic in Indonesia.[9]

Indonesia is undoubtedly on the verge of radical change in telecommunications policy, and in many ways its laws are more clearly established than its neighbours. But until now their implementation has remained uncertain and fragmented. Successful development in Indonesia would certainly have policy reverberations around the region.

Malaysia

With a GNP per capita of US$2,520 Malaysia is, after Singapore, the most developed ASEAN country. With a population of under 19 million, Malaysia has over 2 million telephone subscribers, giving a teledensity of almost 12 per cent.

As in other ASEAN countries, the distribution of telephone lines favours the capital city, Kuala Lumpur and the central region around it, as the following table illustrates. Penang, Malaysia's second city is in the northern region, and Johor Bahru, the third city, is in the south bordering Singapore. Along the east coast of Peninsular Malaysia there are few large towns. In Sarawak and Sabah in Eastern Malaysia (Borneo), population densities are even lower, and still many native people live in villages, or kampongs, strung out along the rivers of dense rainforest. In Peninsular Malaysia people of Chinese and Indian descent tend to live in the urban areas, while bumiputera ('people of the soil') or Malay, Malay-related and aboriginal people dominate the rural areas, while in eastern Malaysia all ethnic groups are represented in the rural areas. See Table 3.3.

Table 3.3
Telephone Data 1992

Region	Subscribers (000s)			Growth % 1988–91	Teledensity Percentage
	Business	Residential	Total		
North	118	380	498	64%	11.0
Central	246	511	757	67%	21.4
South	93	329	422	87%	12.6
East	40	121	161	75%	5.3
Sarawak	50	93	143	57%	8.5
Sabah	39	71	110	34%	6.4
Malaysia	586	1,505	2,091	68%	11.6

Until the licensing of Time Telecommunications (part of the Time Engineering Group) in 1993, Telekom Malaysia was the only supplier of fixed-wireline services. There is no telecommunications monopoly under Malaysian law. Under the Telecommunications Act 1950 (amended several times) the Telecommunications Authority is the Director-General of Jabatan

Telekom Malaysia or JTM (the Department of Telecommunications) under the METP, the Ministry of Energy, Telecommunications and Posts. In 1984, Syarikat Telekom Malaysia (STM) was incorporated and JTM's operations of the national network transferred to it. STM was privatized in 1987 and after flotation in November 1990 as Telekom Malaysia Bhd shareholding was set at 76 per cent Ministry of Finance, 16 per cent local shareholding and overseas investors were initially permitted 8 per cent.

Investment laws in Malaysia are largely driven by the government's Vision 2020 which aims to bring Malaysia up to developed country status by that date. Vision 2020 runs through a series of national plans, including the Sixth Malaysian Plan (1991–96), the National Development Policy (NDP), and the ten year Second Outline Perspective Plan (OPP2) based on the NDP. These plans give a set target for Malaysia of 45 mainlines per 100 people by 2005, an increase of over 9 million lines. The target increase by 1999 is an additional 4 million lines.

But the idea of a planning process is misleading. Up to the present, plan targets have been no more than statements of intent, with no business plans that stretch that far ahead. One reason is that several private companies, with close links to the ruling UMNO (United Malays National Organization) political party, have been given licenses to operate non-basic telecommunications services. Neither they nor Telekom Malaysia know for certain how widely their licences will be extended to compete with Telekom Malaysia. The licensing process itself is opaque, with no obvious criteria being employed, but friends and relatives of leading political figures are often involved.[10] Even when a licence to operate a service has been granted, it often takes several months before any official announcement is made by the METP.

The main burden of responsibility of Telekom is to ensure the provision of rural services, and in this area Telekom is experimenting with the use of radio telecommunications. Telekom is also following the trend of Asian telecommunications companies in looking for business within the region, which includes building contracts in India and the search for business opportunities in China, Vietnam, Indonesia and Bangladesh. Telekom's largest shareholder is PNB (Permodalan Nasional Berhad), Malaysia's largest investment company. Since 1989, PNB has been requested by the government to organize and finance the computer networking of the country's post offices and from there it has extended its networking to establishments such as education offices and road transport offices. PNB owns the majority of container truck companies in Malaysia and it has subsequently branched out to offer similar 19.2 Kbt/s data speed services, seemingly at well below market rates, to some of the largest private companies in Malaysia. Besides the PNB, a number of other private networks exist. Around 15 banks have their own networks for data transport and ATM operations, leasing circuits from Telekom, and Esso, Shell and Petronas have optical fibre networks serving the oil fields. Some national organizations, such as energy, highways, railways and the armed forces also have their own networks.

The spread of computers in Malaysia is occurring as prices of hardware and software fall and this is slowly giving rise to a more information technology-intensive society. As in comparable Asian economies, Malaysia began to focus upon the need to develop a capability in science and technology, including information technology, during the early 1980s. Lowe (1994) records that the Malaysian Administration Modernization and Manpower Planning Unit, which began from a drive by the government to automate its own data processing functions, was elevated in 1985 to a national-level committee responsible for national computer technology policies. At the same time, the Malaysian Institute of Microelectronic Systems (MIMOS) was created as a unit in the Prime Minister's Office to promote a National Microelectronics Programme.[11]

In 1986, Malaysia launched an Industrial Master Plan, which included the electronics and computer technology sectors in its priority list. It was aimed, says Anuwar (1992, p.32) 'at refocusing industrial planning from a largely market-oriented approach to a distinctly planned or target-oriented approach within a free enterprise economy'. These national plan directives were transmitted to Telekom's pressure points. For example, according to Lowe (p.125), at a seminar in 1988 on computerization for development, Telekom was encouraged to 'speed up implementation of an ISDN model' (see Venugopal, 1992). Telekom complains that it is required to undertake these national strategic investments, along with universal service obligations, while remaining uncertain as to what future competition it will face, and without the right to introduce new services of its own, on a commercial footing, without first receiving the authorization of the JTM and METP.

New entrants to the telecommunications market are all local companies. Time Engineering Sdn Bhd, part of the Renong Group, previously the industrial arm of UMNO, is laying optical fibre cable along the federal highways, and its subsidiary company Time Telecom has a licence to offer data, video and audio, and leased circuits to compete with VADS, the joint venture between Telekom Malaysia and IBM which offers value-added data network services to the corporate sector. VADS also offers EDI in competition with EDI Malaysia, now also a member of the Time Group and the provider of EDI services in Port Klang in cooperation with Singapore Network Services. Time strengthened its data network capabilties by being allowed to absorb INC (Information Networking Corporation) which provides user dial-up international online information services by Vsat, using AsiaSat. Time Telecom also claims success in its lobby for a voice licence.

A new entrepreneurial company is Technology Resource Industries Sdn Bhd (TRI) led by Tajudin Ramli who has close personal ties with the Renong Group. In competition with Telekom Malaysia's antiquated ATUR (Automatic Telephone-Using Radio) 450 MHz Nordic system, TRI operates Celcom

(Cellular Communications Network Malaysia), a much more successful ART (Automatic Radio Telephone) 900 ETACS cellular mobile telephone network, and has also been awarded an international telecommunications gateway licence. In 1993, a third cellular licence was awarded to a joint venture, which includes Telekom Malaysia, named Mobikom. This licence covers AMPS and Digital-AMPS and was cutover during 1994 along the Klang Valley, covering the industrial and residential districts stretching from Kuala Lumpur to the port of Klang. TRI is also pursuing markets in Indochina and China[12] and Tajudin Ramli has taken a personal stake in Rimsat, a US-based company which leases Russian Gorizont satellites for orbital slots owned by the Pacific island state of Tonga.[13] In addition, TRI has joined forces with Teleglobe Inc. of Canada to purchase 50 per cent equity in the US Orbcomm project, which is one of several consortia proposing to launch a global LEO (Low Earth-Orbiting) satellite system to provide worldwide mobile and personal telecommunications.

Binariang Sdn Bhd, owned by T. Ananda Khrishnan, a close associate of Prime Minister Datuk Seri Mahathir Mohamad, has been awarded a licence to launch and operate Malaysia's first satellite, Measat. In additional to broadcasting rights, Measat will also offer telecommunications services, including an international gateway. Binariang has also been awarded Malaysia's digital cellular GSM licence. Uniphone, a subsidiary of Sapura Holdings, has the licence to provide urban public call boxes in a revenue-sharing arrangement with Telekom Malaysia. Shamsuddin, Sapura's Chairman, is another associate of the Prime Minister.[14]

Clearly if more voice licenses are issued, and these emerging alternatives can be made to interconnect, Malaysia may end up with two or three competing networks. But as new licences are issued, for example a trail AMPS cellular licence to STW (Syarikat Telefon Wireless) and a PCN licence to Malaysian Media Resources, both companies with strong UMNO ties, the problems of network quality, network management, compatibility of standards, regulation of revenue-sharing and interconnection agreements, will multiply. Although the METP has for long been discussing the idea of a National Telecommunications Policy, and issued a policy statement in May 1994, little of substance has been achieved in the regulatory framework to date. Until it does, it will remain unclear whether a coherent and coordinated development of telecommunications will emerge.

Brunei

Brunei is an anomaly. The country survives on oil. In 1963, the then Sultan opted to keep Brunei Darussalam out of the post-colonial Malaysian Confederation as a measure to avoid sharing this natural resource. The Islamic Sultanate is ruled under a state ideology known as Melayu Islam Beraja (Malay Islam Monarchy, or MIB) which corresponds to Pancasila in Indonesia (see above). The country is sandwiched into two parcels of territory by surrounding Sarawak in eastern Malaysia. In a territory of just over 6,000 km^2, a population of 300,000 (78 per cent Malay and 18 per cent Chinese) live mostly in the capital city of Bandar Seri Begawan. Oil gives the country one of the highest per capita incomes in the world, subsidizes medical and social services, and establishes the Sultan of Brunei as one of the world's richest men.

The telecommunications administration, the Jabatan Telekom Brunei (JTB), thus has no problems in buying any modern equipment it cares to purchase, and is restrained only by a relatively small market and a lack of incentive to build the market. For that reason, and because the base of Brunei's economy is narrow, and income distribution is skewed, telephone penetration is retarded. In 1992, there were only 18 mainlines per 100 inhabitants, and although annual line growth averaged 12 per cent, according to the ITU (1994) only 80 per cent of demand was satisfied, and waiting time for a telephone was still two years. The JTB has plans to raise telephone density to 40 per cent by the year 2000.

As a member state of ASEAN, Brunei has joined in the spirit of forming growth triangles, being part of an East ASEAN Growth Triangle encompassing Indonesia's provinces of Sulawesi and Kalimantan, the eastern Malaysia states of Sabah and Sarawak, and the Philippines island of Mindanao. Tourism, trade and energy resources are considered to be the three legs to this stool.[15]

The Philippines

The Philippine Long Distance Telephone Company, or PLDT, totally dominates the country's domestic and international voice traffic. Until his overthrow in 1986, reports abounded of President Marcos and his cronies using the PLDT to line their pockets, leaving the country with a worsening telephone shortage and a growing debt problem which pushed up prices.[16] Since 1986, the teledensity has improved from 0.8 mainlines per 100 people to 1.5 by the end of 1993. The population is around 65 million.

More than 80 per cent of telephones are in the metropolis, Metro Manila, while over 50 per cent of the country's 1,600 municipalities are not served at all. Outside Manila and the major cities, service is left in the hands of around 60 private operating companies (represented by the Philippines Association of Private Telephone Companies, or PAPTELCO) which are usually confined to one locality and have to give up the bulk of their revenue to the PLDT for long-distance call connections.

As an ex-American colony, the Philippines has inherited an inappropriate constitution, which divides powers between the congress, the judiciary and an executive government. Such a system could work well in the presence of strong institutions of civil society in which the people have, as a right, affordable and equal access to an independent system of legal representation. To the contrary, the people of the Philippines have suffered a society in which corruption is endemic, and where giving bribes has long been considered a cost of doing business. Only since 1986 has the executive government begun to clean itself.

Telecommunications companies wanting a franchise to operate have to secure the passage of a bill through congress. Having gained that, the telecommunications regulator, the National Telecommunications Commission (NTC), must issue a Provisional Authority or PA. This is followed by a public hearing process, which can take years if the PLDT or another local operator raises objections. Finally the NTC must issue a Certificate of Public Convenience and Necessity (CPCN) before a company can offer service without further challenge in the area prescribed by the franchise and the PA. But all is not yet clear. The NTC has quasi-judicial status, and its licences and certificates can be challenged in the law courts, which the PLDT is quick to do whenever it sees competition arising.[17]

President Ramos has joined the ASEAN fashion by announcing Vision 2000 for the development of his country. In pursuit of this Vision he started his term in office like a new broom which brushes clean. Early in 1993 he replaced the Cojuangco family directors of the PLDT; he also cut through red tape by issuing Executive Order 59 in February which mandated the PLDT to interconnect with other operators, and in July issued Executive Order 109 which imposed upon the three holders of international gateway licences (now increased in number[18]) a responsibility to build a minimum of 300,000 lines within the country, and the four (now five) cellular mobile telephone operators a minimum of 400,000 lines each. This rather unorthodox approach to telecommunications planning has been turned to good use by the NTC which invited the eight companies to parcel out urban and rural areas of the country and each take responsibility for one of these composite zones. The allocation of regions is shown in Table 3.4.

These allocations by the NTC only provide PAs, which can be challenged in public hearings.[19] Piltel, a domestic PSTN franchise holder, is jointly owned by the PLDT and the Cojuangco family, and Philcom, an international gateway franchise holder, also has close connections. The cellular operator Extelcom, also closely associated with the Cojuangco family, opted out of the exercise, preferring to focus upon developing its mobile network. Despite these regional allocations it seems the PLDT and the

Table 3.4
Regional Allocations Proposed by the NTC in 1994

Region	Company
1 (N.W. Luzon) + Manila	Smartcom
11 (N.E. Luzon) + Manila	ETPI
111 (C. Luzon)	Smartcom
1Va (S.E Luzon)	Capwire
1Vb (S.W. Luzon + islands)	Globe
V (N. Visayas) + Manila	ICC
V1 (W. Visayas)	Islacom
V11a (W. Visayas)	Islacom
V11b (W. Visayas)	Islacom
V111 (S. Visayas)	Islacom
1X (W. Mindanao)	Philcom & Piltel
1Xb (W. Islands Mindanao)	Philcom & Piltel
X (N. Mindanao)	Philcom & Piltel
X1a (S. Mindanao)	Philcom & Piltel
X1b (E. Mindanao)	Philcom & Piltel
X11 (S.W. Mindanao) + Manila	Globe

Cojuangco family are well positioned to maintain their dominance of the market.

Even if these PA allocations achieve CPCN status, the position of the NTC does not guarantee effective regulation to enforce rules of interconnection. In fact the NTC as regulator and the Department of Transport and Communications (DOTC) as policy-maker are frequently at loggerheads with each other. Personalities and politics bedevil their constant public feuding. The DOTC has gone on record as proposing even more new entrants.[20] A parallel experiment in liberalization is taking place in the skies. Domestic satellite services need to make a major contribution in providing coverage to the many remote rural areas of the Philippines' archipelago, which consists of more than 7,100 islands along a sea-coast stretching 1,800 km. Around 95 per cent of the population live on eleven of the islands, of which Luzon, Visayas and Mindanao are the dominant ones. Domsat has operated a monopoly on Palapa transponders, leased from Indonesia, and rented to domestic carriers, while Philcomsat has enjoyed the monopoly of being the Philippines' signatory to Intelsat (see Table 3.5). During 1994, the exclusivity of both these markets was pronounced at an end. Any carrier with an international gateway licence now has direct access to all international satellite facilities, while the government and seventeen operators have signed a memorandum of understanding to launch the Philippines' first satellite, 'Agila' (Eagle) by December 1996. The regional pattern of every country wanting to have its own bird, and one to spare, intensifies.[21]

The targets of the National Telecommunications Development Plan, drawn up by the DOTC in 1990 to cover the years 1991–2010, originally aimed for a teledensity of 3.5 per cent by 2010. But Vision 2000 requires something more, so rather remarkably this figure has been brought forward to 1997, with the PLDT suddenly willing to become more ambitious and efficient. Adding in the bi-lateral assistance from Japan (in Luzon), France (in Visayas) and Italy (in Mindinao), and rural projects funded by the World Bank and other agencies, the Philippines is, on paper at least, promised an additional 6 million lines by 2001, raising teledensity to 9 per cent. Raising the finance for this expansion is a question yet to be tackled.

In many ways the Philippines is the most open and free of all the ASEAN markets, as Table 3.5 illustrates. There are two dozen companies offering services from paging and cellular mobile telephone to satellite and cable, domestic and international data.

But mostly these companies are controlled by one or other of a small circle of ruling families, usually part of the landed oligarchy. To gain financial and technical backing, or just consultancy, they have alliances with overseas communications companies, including from other Asian countries. Table 3.6 illustrates some of these alliances, which often shift, as they stood in late 1994.

An outstanding feature of these alliances has been their volatility. For example, John Gokongwei's Digitel was partnered by Cable & Wireless until the British company considered further investment too risky. Telstra of Australia was partner to the Lopez family holding company, Benpres Holdings, but also withdrew 'because of "prominent difference" in their appraisal of the business risks in the Philippines project.' (*Telenews Asia*, 6 October 1994, p.3) The Lopez family holdings, which include the regional operating companies Evtelco (Eastern Visayas Telephone Company) and Natelco (Naga Telephone Company) and Gokongwei interests were combined in 1994 under a holding company, Telecommunications Holdings Corporation (THC). The other shareholding will be John Gokongwei's own, through his conglomerate JG Summit, which has also teamed up with Jasmine of Thailand to build optical fibre submarine cables linking the country's major islands. TelecomAsia, the Thai/US alliance of Charoen Pokphand and Nynex, have chosen to fill the vacancy left by Telstra.

Another example of foreign participation is the alliance announced in 1994 between TSI, the major shareholder of PT&T, Capwire and Pocketbell, the A2 Telecommunications Group and Korea Telecom. Through their joint venture, Republic Telecommunications (Retelcom), Korea Telecom will assist with various construction projects, including provision for telecommunications services at the Clark Economic Zone, the former US military airbase at Subic Bay. An example of partial acquisition is the purchase of 16.8 per cent of the shares of Philcom, the international and domestic operator, by Comsat Ventures, a subsidiary of the US-based satellite company and Intelsat signatory Comsat Corporation. The opening of the markets

Table 3.5
Service Operators in the Philippines, 1994

Service	Operators
Domestic PSTN	PLDT (Philippine Long-Distance Telephone Company) PAPTELCO (Philippines Association of Private Telephone Companies — about 60 companies) Digitel (Digital Telecoms Philippines) ETPI (Eastern Telecom Philippines Inc.) Extelcom (Express Telecom Company Inc.) Globe Telecom ICC (International Communications Corp.) Piltel (Pilipino Telephone Corporation) PT&T (Pilipino Telephone Corporation)
Domestic Long-Distance	PLDT (Philippine Long-Distance Telephone Company) Digitel (Digital Telecoms Philippines) Globe Telecom ICC (International Communications Corp.) RCPI (Radio Comms. of the Philippines Inc.)
Inter'l PSTN	PLDT (Philippine Long-Distance Telephone Company) Capwire (Capitol Wireless Inc.) ETPI (Eastern Telecom Philippines Inc.) Globe Telecom ICC (International Communications Corp.) Islacom (Isla Communications Corp.) Philcom (Philippines Global Comms. Inc.)
Domestic Record Carriers	PT&T (PhilippineTelegraph & Telephone) RCPI (Radio Comms. of the Philippines Inc.) TELOF (Telecoms Office — Government) BFC Communications TSI (Telectronics Systems Inc.) UTS (Universal Telecoms Services)
Domestic Satellite	Domsat (Domestic Satellite Philippines Inc.)
Intern'l Satellite	Philcom Sat (Philippines Comms. Satellite Inc.)
Intern'l Record Carrier	Capwire (Capitol Wireless Inc.) ETPI (Eastern Telecom Philippines Inc.) Globe Telecom Philcom (Philippines Global Comms. Inc.)
Mobile Cellular	Piltel (Pilipino Telephone Corporation) Extelcom (Express Telecom Company Inc.) Globe Telecom Islacom (Isla Communications Corp.) Smartcom (Smart Information Technologies Inc.)
Paging	Easycall/Matrix (Australia), Pocketbell (Santiago) Beeper 150, Digipage and PowerPage.

Note: Piltel is a subsidiary of PLDT.

Table 3.6
The Families and Partners

Family	Company	Foreign Participant
Ayalas	Globe Telecom	Singapore Telecom Intn.
Delgados	Isla Communication	US West & Shinawatra
Gokongwei	Digitel	Telecom New Zealand
Lopez	ICC; RCPI; Evtelco; Natelco	Nynex/TelecomAsia
Ortigas	Beltel	Bell Atlantic[1]
Osmena	Smartcom	First Pacific (HK)
Razon	TTPI/ETPI	Cable & Wireless
Santiagos	PT&T/Capitol/TSI	

[1] Not formalized as of 1994.

decreed in 1993 has stimulated growing foreign interest, but equally a lot of jockeying for licences, for alliances and for finances.

The Constitution of 1987 expressly limits foreign investment in public utilities to 40 per cent, and the Finance Investments Act 1991 places telecommunications on a Negative List which removes discretion from the Board of Investment to waiver the rule. However, under the 'grandfather rule' foreign companies can also own up to 40 per cent of the local joint venture partner, giving the foreign partner a maximum 64 per cent share of the dividends. Also an amendment in 1994 to the 1990 Republic Act No.6957, which institutionalized the Build-Operate-Transfer (BOT) concept for public sector infrastructure projects, extended BOTs to include information technology networks and database infrastructure. In the telecommunications sector, unlike other sectors, there is no shortage of interest by foreign capital, but equally no shortage of problems to be resolved before hard cash is sunk in the ground.

Besides the crippling problem of having to service a huge foreign debt inherited from the Marcos years, the Philippines economy suffers continuously from power shortages. However, there is hope that in the near future this problem will recede, and then the economy has chances. But the Philippines suffers dreadfully from distorting social inequalities, disabling poverty (the GNP per capita is US$730)[22], and political graft and violence which have rotted the roots of development. Now, for the first time, there seems to be a real opportunity for small towns and villages across the Philippines to gain access to a telephone, but at the same time this could turn out to be yet just another opportunity for local officials to exploit the business.[23] Development — social and economic — without telecommunications is not possible; but neither is telecommunications in a country without development. It will take something more than a few telephone wires to break out of this vicious circle.

Thailand

There are, in effect, competing telecommunications authorities in Thailand. The Telegraph and Telephone Act 1934, and the Radio Communications Act 1955 (which superseded six previous acts) gives the PTD (Posts and Telegraph Department) of the Ministry of Transport and Communications, or MOTC, rights over radio telecommunications, but the PTD's authority over domestic fixed-wire telecommunications was seceded to the TOT under the Telephone Organization of Thailand Act 1954, and over international telecommunications to the CAT, or Communications Authority of Thailand, under the Telecommunications Authority Act 1976.

Originally the TOT was exclusively responsible for domestic public switched voice traffic, including 'trunk' calls to Malaysia, Cambodia and Laos, and the CAT for international services. In practice, they are emerging as competitors in four areas: cellular telephone, paging, data networks and leased circuits, reflecting major changes now taking place in the country's telecommunications sector. Furthermore, advances in technology will erode the functional separation of fixed-wire and radio systems, as for example in the use of radio in the local loop, and radio-links for rural PSTN and mobile data-over-voice systems, and this in turn will pitch the PTD's powers of issuing licences into competition with both the CAT and the TOT.

Since the mid-1980s, Thailand's rate of growth of mainlines has averaged 16 per cent per annum, driven by a foreign investment-led boom of the Thai economy. See Pupphavesa and Stifel (1993). But Thailand's population of 59 million still has only 3.3 telephone lines per 100 people. With a GNP per capita of US$1,570 Thailand should, by ITU averages, have over 4, but it has an unusually large proportion of mobile subscribers, around 14 per cent.

Thailand's rapid economic growth rates — per capita income rose annually by 15 per cent from the mid-1980s — has put enormous strains on the country's communications infrastructure, not least telecommunications. Under existing Thai law, public networks belong to the state. After the restoration of democracy in 1992, following the military coup of 1991, one of the first steps of the new government was to remove the influence of the air force and army respectively from the policy-making of the CAT and TOT, and then propose plans to push ahead with the privatization of both the CAT and the TOT. The growing pressure to expand the network and improve the quality of services, to increase investment, is forcing change.

The first shift of gear came in 1990 with the introduction of BTOs (Build-Transfer-Operate) which allowed private companies to build out value-added networks, transfer the ownership, but operate the service through revenue-sharing arrangements with the TOT and CAT. Two private companies, Shinawatra and United Communications (UCOM) have cellular mobile BTOs with the TOT and CAT, in addition to the TOT's and CAT's own cellular services. In paging, there are four BTOs, two each with TOT and CAT. In the Vsat market for data communications the TOT and CAT have one BTO apiece, but two other private companies have separate licences issued by the PTD which regulates the use of radio frequency for the MOTC. In many of these BTOs, the local Thai company also has a foreign partner.

The second change, agreed in 1992, was to license two consortia to build two million telephone lines in Bangkok and one million in the provinces. TelecomAsia, a joint venture between Charoen Pokphand and the American Bell company Nynex, has the Bangkok BTO and TT&T (Thailand Telephone and Telecom), a joint venture led by Loxely and Jasmine,[24] the provincial BTO. Jasmine, in partnership with Northern Telecom of Canada, also has a BTO with the TOT to lay a submarine optical fibre cable across the Gulf of Thailand, and the local Comlink Group has a similar BTO to lay a cable along the state railway tracks.

True to all developing Asian economies, local companies that win telecommunications concessions, franchises and licences, must have close personal, military and political connections. The most successful to date is Dr Thaksin Shinawatra,[25] an entrepreneur who holds an American doctorate in criminal justice, and was deputy head of the Thai police computer centre until he turned to the business of procuring IBM computer systems for the police, the military and government departments. Besides cellular mobile, paging and broadcasting licenses, his Shinawatra company won, in 1990, the right to launch and operate Thailand's first satellite, Thaicom-1. Originally the concession granted by the civilian government of Chatichai Choonhavan was exclusive for thirty years, but one military coup and civilian government later, in 1992, this was revised to eight years exclusivity.[26]

According to *The Asian Wall Street Journal*, 26 November 1993, Shinawatra had a similar local difficulty in Cambodia when he negotiated a 99 year broadcasting concession with the ruling Cambodian People's Party, the CPP. The awarding of the concession was subsequently investigated for 'irregularities' after the CPP lost the election in May 1993. However, Shinawatra has gone on to win contracts for installing telephone lines and wireless local loop in Phnom Penh, fixed wire, mobile and broadcast concessions in Laos, and paging in Vietnam. China and Burma are the next markets Shinawatra is looking to enter, having succeeded in entering the Philippines cellular (and now fixed-wire) market by taking a stake in Isla Communications.

The Sino-Thai Chearavanont family (which owns Charoen Pokphand, Thailand's leading agri-business conglomerate) and TelecomAsia are equally well connected. The original BTO contract awarded to TelecomAsia by the TOT during the government of Chatichai Choonhavan was for 3 million lines countrywide with a concession for 25 years on very favourable terms. After the fall of the government this contract was also reviewed by the caretaker Prime Minister Anand Panyarachun, who managed to maintain a reputation for clean government despite conflicts with members of the military junta who favoured the deal. Under the revised terms Charoen Pokphand/TelecomAsia was confined to 2 million lines in Bangkok.[27] Without its BTO concession, the company would have no telecommunications business at all, yet now TelecomAsia commands around 10 per cent of the Bangkok stock market, and claims building contracts in Hubei Province in China, and in Vietnam as well as Cambodia. TelecomAsia also has an agreement with the TOT and the Mass Communications Organization of Thailand (MCOT) to construct a cable television network in Bangkok, and is proposing to offer a videotex service with Lines Technology, which offers a similar service in Singapore.

UCOM is another family business owned by the Bencharongful family, which, like Shinawatra, began as a supplier of mostly radio-based communications equipment to the police, military and government. As Shinawatra teamed up with IBM, so in 1969 UCOM teamed up with Motorola, and UCOM is now Thailand's signatory partner in Motorola's worldwide Iridium mobile telephone satellite project. Motorola holds 15 per cent of UCOM stock. UCOM is also the junior partner with Shinawatra in Thailand's CT2 network, along with Singapore Telecom International, a minority shareholder in Thaicom-1, and in December 1993, jointly with Shinawatra, signed a contract with TOT for the construction of a fixed-wire and wireless local loop network to serve rural areas.[28] The close relations between Thai companies is further confirmed in UCOM's case with TelecomAsia holding 6.5 per cent of its shares. UCOM's foreign ventures are currently confined to sub-contract network construction, for example for Chareon Pokphand in Cambodia.

The other Thai communications company with overseas investment is Samart, which operates the largest of the cellular networks in Cambodia. Like Shinawatra and UCOM, Samart has for long been a supplier of corporate and military communications equipment.[29] The linkages between these companies and the Thai business-military complex is not surprising since they offer a protected or privileged home market. But they also correspond to a wider political economy, the drive within Thai business, military and political societies to become the gateway to Indochina, not to say the gatekeeper. At one end of the scale is Thai domination of the region's markets, at the other is Thai containment of the political spheres of influence of China, India and Vietnam. Expressed as a telecommunications issue, Thailand wants to become Indochina's communications hub for phone, fax, data and broadcasting.

In the absence of a stable central government the aspirations of the various business, military and political groups to promote a Thai regional hegemony fails to gain coherent expression. But democratic counter-thrusts are also too weak to establish a clear perspective for the development of the country, so Thailand drifts along with a highly imperfect but 'free' market, economically attractive to foreign investors, yet with effective planning by government paralysed. The result is infrastructure chaos.

The five-year national economic and social development plans, drawn up by the National Economic and Social Development Council (NESDC) and approved by the Prime Minister's Office and Cabinet, saw no need to change the status of the telecommunications industry. But the boom of the 1980s exposed the chronic shortage of telephones and pushed the government towards opening up through the BTO scheme. The advantage of the BTO, that is the transfer of ownership after the 'build' stage but before the 'operate'stage, was that it allowed circumvention of telecommunications laws, which require basic services to remain under state control. But the Five-Year Plan (1992–97), which targets line growth at 3 million and a teledensity of 9.7 per cent by 1997,[30] for the first time envisages the need for Thailand to go beyond the resources of the TOT and CAT. Both the TOT and CAT are to be fully corporatized, and partially privatized. The Eighth Five-Year Plan (1997–2002) foresees annual line growth of 1.2 million.

It is proposed to lift restrictions on foreign direct investment, currently limited to 40 per cent, and create a regulatory body, but all the signs are that this new body — unofficially known as the National Telecommunication Commission — will consist of far too many vested interests, presided over by the minister, rather than a semi-autonomous body of independent specialists. In practice, the PTD may emerge as the effective secretariat of such as body. But changing the laws will require a government to be in office long enough to do it.

For the reasons just given, the vision of Thailand becoming a telecommunications hub is far-fetched at this stage. The need is recognized to develop not just the basic network, but also Thailand's information technology infrastructure. In 1992 an Information Technology Development Committee was established to advise government, and the TOT has ambitious plans to make ISDN services available to corporate customers in Bangkok on a commercial basis by 1997, and available to small business users by 2000, and broadband services available thereafter.

In another initiative, in January 1994, Thailand's first teleport opened next to the Intelsat earth station at Laem Chavang, close to the Eastern Seaboard industrial development project. The CAT and TOT will cooperate to divert heavy traffic away from Bangkok, but criticism that the nearby deepwater sea port is not deep enough raises doubts about the success of the teleport in attracting companies to locate. A further project, that of a Data Processing Zone, has never got off the ground, partly because no-one could decide whether the zone was spatially physical or virtual.

But private companies, which are allowed into the value-added services sectors, are starting to offer on-line services. For example, IBM's

Information Network offers international networking in partnership with the CAT, and since January 1994 so does the International Trade Information (Thailand) Co. The domestic Vsat operators (see Table 3.7) also offer data networking. It seems that Thailand's telecommunications future will lie more in liberalization, given the impossibility of planning, than in planning the impossible.

Table 3.7
Build-Transfer-Operate

	Operator	Service				BTO Network
Telephone Lines	TelecomAsia	2 million lines in Bangkok				TOT
	TT&T	1 million lines in Provinces				TOT
Payphones	AIS	Cardphone				TOT
Cellular	CAT	Analogue	AMPS	800A	Mhz	
	TACS (UCOM)	Analogue	AMPS	80-0b	Mhz	CAT
	TACS (UCOM)	Digital	DCS	1.8b	Ghz	CAT
	TOT	Analogue	NMT	470	MHz	
	AIS (Shin)	Analogue	NMT	900E	MHz	TOT
	AIS (Shin)	Digital	GSM	900	MHz	TOT
CT2	Shinawatra	Fonepoint				TOT
Paging	Pac. Telesis	Paclink				CAT
	Matrix	Easycall				CAT
	Percom					CAT
	Shinawatra	Phonelink				TOT
	Loxley	Pagephone				TOT
Trunked Radio	UCOM					CAT
	Acumen					TOT
Data	CAT	PSDN				
	ThaiSkycom	Vsat				CAT
	TOT	PSDN				
	Shinawatra	PSDN				TOT
	Acumen	Vsat				TOT
	Samart Telecom	Vsat				PTD
	Compunet	Vsat				PTD
Videotex	Lines Technology/TelecomAsia					TOT
Long-Distance	Comlink	Along railway tracks				TOT
Optical Fibre	Jasmine	Submarine cable				TOT

Note: TelecomAsia is owned by Charoen Pokphand (Nynex and Orient Telecom are partners); TT&T (Thailand Telephone & Telecom) is jointly owned by Loxley/Jasmine (Ital-Thai and Phatra Thanakit are partners); AIS (Advanced Information Services) is a subsidiary of Shinawatra; TACS (Total Access Communications) is a subsidiary of UCOM (United Communications Industry Ltd). Acumen is a subsidiary of Jasmine. Several foreign companies are partners to Thai firms, for example Cable & Wireless with Srifeungfung Sophonechich in Compunet, France Telecom/Matra (owns Lines Technology) with TelecomAsia in Videotex; Hutchison with Loxely in paging; Nynex with Charoen Pokphand in TelecomAsia, STI (Singapore Telecom International) with Shinawatra and UCOM in CT2, paging and data, Telstra with Samart Telecom.

Vietnam

After years of war, and peace since 1975, only 350,000 telephone lines were in service by mid-1994 for Vietnam's 71 million population, 80 per cent of whom live in rural areas. Telephone exchanges had been installed in 350 of the 527 districts, most of the lines serving government offices, the military and businesses. Residential lines are no more than 10 per cent of the total in a country where GDP per capita in 1992 was only US$100.[31]

Economic reconstruction is the priority of the Vietnamese Communist Party and government, which replaced the Stalinist constitution of 1980 with a new constitution in April 1992 in which the guiding principle is economic innovation or 'doi moi'. Central to the reform was the dropping of direct state control of foreign economic relations, and writing into the constitution the Foreign Investment Law which permits joint ventures, and 100 per cent foreign ownership of assets. But in the sensitive area of telecommunications, still regarded as a national security issue, [32] the only form of foreign participation permitted is the Business Cooperation Contract, or BCC, which is an agreement between a foreign and a Vietnamese partner for 'the mutual allocation of responsibilities and sharing of product, production or losses, without creating a joint venture enterprise or any other legal entity.' See DOTC (1993, p.11, fn 7).

Vietnam has to rely heavily upon foreign assistance for the renovation of its telecommunications sector. Direct telephone connections to the USA were re-opened in 1992, and the lifting of the US trade embargo in 1993 has finally opened the doors of the World Bank, the Asian Development Bank, the International Monetary Fund, and other multilateral agencies, all of which are necessary to support the convertibility of the Vietnamese currency (dong). Paying for imported telecommunications equipment is also helped by vendor credits, but this threatens to overwhelm Vietnam with too many different and incompatible types of switching and transmission equipment. Vendor credits include French, German, Italian, Korean and Swedish companies.[33]

Vietnam's first step towards renovation came in 1986 when the Directorate-General of Posts & Telecommunications, or DGPT, negotiated a BCC (started 1988) with Telstra — previously OTC and now part of Telecom Australia — to build a series of satellite earth stations for international communications with Vietnam's extensive overseas community. Standard A Intelsat earth stations have been built in Hanoi in the north and Ho Chi Minh City (HCM City) in the south[34] and a smaller Standard B at Danang in central Vietnam. International revenues, which are shared between the DGPT and Telstra, are vital to the DGPT's plans to finance domestic network expansion. Telstra, which is extending its involvement to domestic network planning and consultancy, is also financing the DGPT's share of an optical fibre submarine cable linking Vietnam to Thailand and Hong Kong due for completion by the end of 1995. Microwave links to neighbouring Cambodia and Laos are also being set up.

In 1989 a *Strategic Plan* for network expansion was drawn up with the aim of raising mainlines from 120,000 in 1988 (a penetration of 0.2 per cent) to 370,000 (or 0.5 per cent) by 1995, and to 867,000 lines (or 1 per cent) by 2004. However in 1993 (see *Asian Communications*, November 1993) these targets were revised upwards to read: 700,000 (or 1 per cent) by 1995, 1.6 million (or 2 per cent) by 2000, and 4 million (or 5 per cent) by 2005. By the end of 1993 HCM City, the commercial centre, had around 90,000 lines, a penetration rate of about 2 per cent. It aims for 200,000 lines by 1995 and 500,000 by 2000. These targets are ambitious given that most investment in recent years has been in equipment replacement rather than line growth, but Vietnam is set on building from new, which means a fully digital network, and aims to be self-financing by 2000.

Starting with Hanoi and its port-city Haiphong, and HCM City, Vietnam has now fully digitalized its small but growing network, including the national trunk network and the three international gateways. Optical fibre links between the main cities, and microwave links to the major towns of the 53 provinces are being constructed, and further microwave links are being used to connect rural areas, but so far only one quarter of Vietnam's villages have telephone links, although the target penetration is 70 per cent by 2000. Telex and telegrams are still widely used across Indochina, and Vietnam has automated these for international traffic. At the other end of the scale, a public data network is available in Vietnam providing X.28 dial-up (asynchronous mode) and X.25 dedicated access (synchronous mode) transmissions as well as X.400 E-mail. Vietnam's VARENet (Vietnam Academic Research and Education Network) connects with Internet through Hanoi's Institute of Information Technology (IOIT) hubbing through the Computing Unit of the Australia National University. Computers in Vietnam are widely available for those who can afford them, but power outages are a problem.

These services are monopolized, and therefore regulated, by the DGPT, which for a brief period from 1990 to 1992 was part of the Ministry of Communications and Transport. When the ministry was split into two, the DGPT emerged as an independent entity, with the Secretary-General receiving ministerial rank. A new body, the VNPT, or Vietnam National Posts & Telecommunications, was formed to operate the national network under the regulation of the DGPT. Under the VNPT come: Vietnam Telecoms International (VTI), Vietnam Telecoms National (VTN), Vietnam Mobile Services (VMS), Vietnam Data Corporation (VDC), the Vietnam Postal Service (VPS), and the city and provincial P&T bureaux. The VNPT is also responsible for the provision of radio and television broadcast transmission. The government has indicated that in 1996, following the opening of the Hanoi Stock Exchange in 1995, a public share offering in the VNPT will be made.

The national priority afforded telecommunications is indicated by the Prime Minister's Office having its own advisers, and by the fact that DGPT tariffs require cabinet approval. As yet no clear distinction has been drawn between basic and value-added services, but in practice radio and data communications have been opened to foreign participation as 'trials' which could lead to BCCs, as the following table illustrates.

Table 3.8
Foreign Participation in Vietnam's Telecommunications

Service	Operators	Location
Cellular	Singapore Telecom International	HCM City
CT2	Steamers (Keppel Corp, Singapore)	HCM City & Hanoi
Paging	ABC (Hong Kong)	HCM City & Hanoi
	Epro (Hong Kong)	HCM City
	MCC (Australia)	HCM City
	Shinawatra (Thailand) with	HCM City
	Singapore Telecom International	HCM City
Payphones	Sapura (Malaysia)	HCM City & Hanoi
	Schlumberger Technologies(France)	Hanoi
Trunk Radio	Steamers (Keppel Corp, Singapore)	HCM City
Packet-Switched Data	Telstra (Australia)	HMC City & Hanoi

The DGPT has authorized the go-ahead for GSM cellular systems in Hanoi (Alcatel equipment) and HCM City (Ericsson equipment) casting doubt over the future of the STI 'trial' AMPS system in HCM City. A nationwide paging service (Motorola equipment) is also planned by VMS. Telstra has been running a packet-switched data service, again on a trial basis, and private companies are encouraged to install their own facilities for interconnect with the public network and sell value-added services.[35]

The need to develop an information technology infrastructure and information services was recognized in 1993 by the issuing of an IT masterplan called ITP-2000. Anticipating the need for IT is a memorandum of understanding between Australia and the Vietnam Trade Information Centre, which is part of the powerful Ministry of Commerce. Australia's Bureau of Transport and Communications observes this 'has the potential to become the basis of an alternative communications network' (DOTC, 1993, p.11). And in a development that has some parallels to China, the Prime Minister's Office has authorized the Ministry of Defence to set up an Army Telecommunications Company (ATC) to provide public network services. The network-building capacity of ministries has two advantages for Vietnam: it keeps networks under state ownership while providing a revenue-base for the cash-strapped administration.

Vietnam is identified by many telecommunications companies as a key growth market in Asia in terms of equipment sales, consultancy contracts, and traffic revenues. Vietnam starts from a lower base than China and with greater funding difficulties, but the DGPT's stated aim is to become self-financing by 2000. An early presence in Indochina has placed Australian companies in a strong position, but with the lifting of the US embargo competition to enter the Vietnam market will become intense, not least from France Telecom for historical reasons.Vietnam will become an interesting case to follow as it steers a course between openness and national control, with strong economic and social development goals.

Burma, Cambodia, Laos

Burma (Myanmar), with a population of 44 million, has a telephone network of 80,000 mainlines recorded by the ITU (1994) for 1992, just 0.18 telephones per 100 inhabitants. Following the military coup in 1962, a policy of economic autarky has left the country isolated, but, as Steinberg (1990) points out 'the world has changed. Burma could be isolated in 1962; it cannot today. Anyone overseas can direct dial a cabinet minister or leader of the opposition.' (So long as the leader is not in detention, of course.) The *Asian Wall Street Journal*, 25 April 1994, reports Chinese engineers have installed 300 new telephone lines in the capital Rangoon (Yangon), and China provides access to AsiaSat (Hong Kong) for international direct dialing.[36]

Myanmar Posts & Telecommunications (MPT) is responsible for the PSTN under the authority of the Ministry of Transport and Communications (MOTC). According to the Five Year Telecommunications Development Plan (1988–92) the network was to have doubled. In fact it rose by just 5,000 lines, or 7 per cent. Only 18 per cent of exchange lines are digital, although one recent addition has been a mobile cellular network installed by Ericsson in the capital.

The country is a natural communications centre, situated at the crossroads to China, India and Thailand. Once its isolationism is ended communications will again flourish. China, India and Thailand for their own reasons are all very interested to see that happen. But it was a xenophobic reaction against those influences that pushed Burma's military dictatorship to cut themselves off in the first place, and by expropriating local Chinese, Indian and Thai traders, reduce the country from riches to rags. Burma's GDP per capita stood at US$565 in 1992.

Cambodia's GDP per capita, with just 9 million population, stood at US$225. Besides the untold losses suffered during the Vietnam War, during the regime of the Khymer Rouge, and the subsequent civil war, Cambodia also faces the uncertain ambitions of Thailand's military. Despite these difficulties, the regime in Phnom Penh manages to offer telecommunications concessions to foreign companies. The Directorate of Posts & Telecommunications (DPT) claims 5,000 mainlines, or 0.06 teledensity.[37] As in Vietnam, Telstra has put in a Standard A earth station for international communications.

Other radio-based foreign concessions involving joint ventures with the DPT are Casacom, with Samart (Thailand) using NMT 900, Camtel, with the CP Group (Thailand) using Motorola 800 AMPS handsets, Tricelcam with Technology Resources (Malaysia) using 900 TACS, and Shincam with Shinawatra (Thailand) which has a concession to install a NMT-450 wireless local loop network in Phnom Penh, as well as to operate a private broadcast station. There is also an Vsat network, supplied by Telstra, and left behind by the United Nations (UNTAC) peace-keeping forces. Cambodia's government has called for bids to install telephone facilities at 20 sites around the country, to be linked by satellite. Ericsson has supplied an AXE switch providing international circuits and 5,000 local lines for Phnom Penh, and Cambodia has negotiated with PT Telkom to use Indonesia's Palapa-B transponder capacity for domestic communications. So within the capital city, and within the military, networks exist, but without peace the issue of telecommunications and development in Cambodia remains academic.

Laos, a rugged and landlocked country sandwiched between Vietnam and Thailand, with China to the north and Cambodia to the south, has a population of only 4.5 million, many of them hill tribes, and 17,000 mainlines, or 0.4 teledensity. GDP per capita in 1992 was US$245. Despite Lao's fears of commercial colonialism by Thailand, the government agreed to relax its borders in 1994 with the opening of the diplomatically named Friendship Bridge across the Mekong River. Trade will increase communications.

Since 1992 the Enterprise d'Etat des Postes et Telecommunications Lao (EPTL) has been running a single-cell AMPS cellular system purchased from Systar Telecoms (USA). Caution about Thailand's intentions did not stop the government in 1993 from granting the Thai company Shinawatra a concession to develop telecommunications services on a joint venture basis covering PSTN, international telecoms, a GSM cellular mobile service for Vientiane, payphones, digital paging, radio and television broadcasting and, according to the company's 1993 'Communications For Better Living' publicity, 'other projects, specified by the Lao government.' The projects are part of Lao's Third National Development Plan.

Desperately poor, remote and landlocked as Laos is, it has peace, and a government that wants to liberalize the economy without losing national identity or sight of its development goals. This has attracted funding from the UNDP and the World Bank (nearly US$5 million in 1991) as part of a modernization programme, and further multilateral and bilateral aid can be expected. If the Open Door can be handled successfully, telecommunications in Laos will have a positive role to play.

---

[1] AFTA formally came into being 1 January 1993, creating a market potential of 330 million people. In 1991, intra-ASEAN trade flows totalled an impressive US$33 billion, or 11 per cent of integrated GDP, but a closer look at the IMF's Directions of Trade Statistics Yearbook shows that 87 per cent of that was bilateral (and re-export) trade with Singapore.

[2] Malaysia in particular has expressed concern that AFTA will be highjacked by an American-led APEC (Asia Pacific Economic Cooperation) and instead has argued for an EAEC (East Asian Economic Caucus) with membership restricted to 'Asia' which would include Japan but exclude Australia, New Zealand and the USA. This is consistent with the 'Look East' policy long espoused by Malaysian Prime Minister Dr Mahathir.

[3] Indonesia's Statistical Bureau (Biro Pusat Statistik) estimates income distribution from per capita expenditure data. Between 1984 and 1990 the bottom 40 per cent of the population increased their share by just 0.56 per cent to 21.31 per cent, while the top 20 per cent maintained their 42 per cent share.

[4] Indonesia's middle-class is estimated at 17 million. See DOTC (1993) p.1.

[5] The deregulation also applies to the media industry, but the Hong Kong *Eastern Express*, 6 June 1994 reported the Information Minister Harmoko as objecting that 'if the Press Law [of 1966] does not allow foreign investment, then this law is still effective. I hold to the Press Law.'

[6] Around 80 per cent of Indonesia's top 200 conglomerates are owned by Chinese. 'That figure alone causes great resentment even among the least prejudiced people . . . Hence opposition to deregulation often coincides with hostility towards all Chinese, making further deregulation an unpopular course to advocate in present circumstances.' Professor J.A.C.Mackie, *Eastern Express*, Hong Kong, 28 April 1994. The Salim Group alone, led by Liem Sioe Liong, probably Asia's richest man after the Sultan of Brunei, is said to account for 5 per cent of Indonesia's GDP. See *Far Eastern Economic Review*, 10 August 1991. As in Malaysia, ethnic Chinese, whatever their personal circumstances, are always vulnerable to being used to deflect more deep-rooted social problems. In April 1994, labour unrest and anti-Chinese rioting broke out in the Indonesian city of Medan where many of the employers are Sino-Indonesians. Radio Televisyen Malaysia (RTM) censored a BBC news bulletin depicting the events on the grounds that it

would offend their neighbour and cause undue concern in Malaysia itself. See *Eastern Express*, Hong Kong, 29 April 1994.

7   Following the political decision to split a controversial telephone equipment contract between America's AT&T and Japan's NEC, local companies associated with the President's eldest daughter, Tutut Siti Hardijanti Hastuti, and two of his sons, Bambang Trihatmodjo and Tommy Hutomo Mabdala Putra, were appointed as the local agents. For an overview of the leading family's interests, which include satellite and GSM mobile telephones, see *Far Eastern Economic Review*, 30 April 1992.

8   The World Bank favours more open markets and better incentives while using a greater share of limited government spending for education to improve skills. In contrast to these suggestions, the bank states, 'policies centred on a "technological leapfrogging" involving the development of targeted high-technology industries supported by direct public investment or subsidies and high levels of protection, have proven costly and ineffective in most countries.' *The Asian Wall Street Journal*, 8 June 1993, p.1.

9   A further example is the proposed East Asian Growth Area (EAGA) of Borneo, comprising Indonesia's Kalimantan province, Sabah and Sarawak in eastern Malaysia, and Brunei. The Philippines, which has never publicly relinquished its claim over Sabah, and possibly feeling left out, has floated the less likely idea of a North Asian Growth Area of Northern Luzon island, Hong Kong, Southern China and Taiwan, and possibly Japan, but following a meeting between President Ramos and Malaysian Prime Minister Mahathir Mohamad in early 1994, the Philippines has also proposed a growth triangle covering Mindanao, Sabah and the Indonesian island of Sulawesi. In a different geography, South Korea has alluded to a growth square covering the North-East of China, South-Eastern Russia, North and South Korea, and Japan.

10 Lowe (1994) points out that there 'is a close relationship between telecommunications firms and ruling elites. However, in Malaysia ethnicity is the principal element in this, while in Indonesia, it is more family connections.' (p.118) This is a reference to the stratification of political parties in Malaysia along ethnic lines and to the strategy of the Malay majority party, UMNO, to promote bumiputera interests. Lowe correctly goes on to emphasize that essentially this is a political process. Political parties need votes. They also need finance and financially strong backers. Money politics plagues many aspects of Malaysian business.

11 According to the MIMOS publicity brochure, undated, circa 1989, 'the nation's economic prosperity is dependent on the success with which we manufacture IT products and exploit IT services.' Among other bodies, including the universities, the influential think-tank MIER (Malaysian Institute of Economic Research) has sponsored numerous government-supported seminars on strategies for IT development and many publications.

12 In a letter to the Kuala Lumpur Stock Exchange, dated 21 September 1993, TRI claimed a joint venture agreement with a People's Liberation Army (PLA) busi-

ness unit, the China Poly Economy Technology Company, to provide turnkey cellular systems.

13. The rapid expansion of Asian stock markets of recent years has offered local entrepreneurs considerable scope for recouping highly speculative investments with press announcements which boost their company share prices. The telecommunications sector has provided many examples of this across the region, and it has become a significant mechanism for entrepreneurial market entry. In cases such as Malaysia, where market entry enjoys a degree of protection, economic risk is balanced by political risk.

14. Sapura's Deputy Chairman, Rameli Musa, is a schoolfriend of Finance Minister Anwar Ibrahim, Dr Mahathir's potential successor. College and personal networks are just as important in Malaysia as they are in North America or Western Europe.

15. Other ASEAN growth triangles include the Southern Growth Triangle (consisting of Indonesia's Riau province, Singapore and Johor Bahru in Malaysia), and the Northern Growth Triangle (consisting of southern Thailand, northern Penninsular Malaysia, and Indonesia's province of Sumatra).

16. For a detailed history of manipulation and control of the PLDT by one branch of the Cojuangco family, close associates of ex-President Marcos, see Manapat (1993) and various issues of the *Far Eastern Economic Review*.

17. A ruling in 1993 against a licence issued to Eastern Telecommunications Philippines Inc (ETPI) — of which Cable & Wireless is a partner — to operate a second international voice gateway in competition with the PLDT caught public attention when the Judgement was deemed, by a handwriting expert, to have been written by the PLDT's own counsel. See Manapat (1993) and *Far Eastern Economic Review*, 11 February 1993. The issue went to appeal. Another company with close PLDT connections, Philcom, was also granted a licence which went unchallenged.

18. By 1994, six international gateway franchises had been approved by the NTC: besides the PLTD, they are Capwire, ETPI, Globe Telecom, ICC, Islacom, and Philcom.

19. A challenge could come from excluded companies, like Beltel and Digitel. Digitel has a particularly strong case given that it bid for and was awarded the right to operate throughout Luzon under the government programme to privatize the networks built with Japanese aid money. Digitel's contract stated that its only competitors would be existing Paptelco operators and the PLDT.

20. A public account of a government-private sector dialogue held under the auspices of the Philipine Electronic and Telephone Communications Foundation (PETEF) in June 1994 reports the DOTC Secretary and the DOTC Undersecretary and the NTC Commissioner all in open disagreement with each other. See *Telenews Asia*, 16 June 1994, p.2.

21. If this is launched, then Indonesia loses a customer, and Malaysia, hoping to gain a customer after the launch of its own Measat, will have to go it alone.

22 One important source of revenue for the Philippines is the repatriation of earnings from overseas migrant workers. Most families in small towns and villages in the Philippines could only afford a telephone with this support.

23 From another angle, Senator Ernesto M. Maceda complained that the NTC's allocation plan threatened to 'open the floodgates of graft and corruption in the warding of profitable areas to selected firms.' *Telenew Asia*, 16 June 1994, p.2.

24 Loxley is owned by the Lamsam family which also runs Thai Farmers Bank, Thailand's third largest bank. Jasmine began as an engineering contractor to TOT, and is working with TOT to enter the Indochina market, and has also teamed up with JG Summit in the Philippines to manufacture and install submarine cable links. (See Philippines section above.)

25 Dr Thaksin Shinawatra was appointed Thailand's Foreign Minister in November 1994. According to a report in the Hong Kong *South China Morning Post*, 4 December 1994, his first act was to debug his government office. His predecessor was the chief of intelligence.

26 In part the pretext for the coup in 1991 was an accusation of corruption by the Chatichai Choonhavan government in the award by the MOTC to Shinawatra. The re-negotiated concession still requires users of other satellites, such as AsiaSat and Palapa, to transfer to Thaicom as soon as their existing contracts expire, or sooner if Shinawatra can buy out the penalty clauses in their contracts.

27 When the deal was finally confirmed in July 1992, the *Far Eastern Economic Review*, 16 July 1992, p.55, quoted Anand Panyarachun as follows: 'State employees acted for their own benefit and abused their authority in helping private concessions through dishonest and non-transparent methods, causing [potentially] heavy losses for the nation.' For details of the saga, which underscores serious transaction cost inefficiencies, see Charmonman (1994).

28 The award was accompanied by now familiar litany of complaints about irregularities in the bidding process. See *Asian Communications*, March 1994. For a study of the social and economic benefits of rural telephony in Indonesia and Thailand, see Chu, Srivisal, Alfian and Supadhiloke (1985).

29 Shinawatra holds a 20 per cent stake in the subsidiary, Samart Satcom, which makes satellite dishes. The linkage between these companies is quite strong, and seems characteristic of the close commercial and personal networking of Asian business communities.

30 By contrast, the TOT estimates demand will be 7 million and teledensity will be 11.43 by 1997. Information by interview.

31 The World Bank uses GNP per capita to classify countries into low, middle and high income economies. The GDP per capita data for Indochina comes from the ITU (1994). In the case of Indochina net overseas income may be significant only for Vietnam, and then probably not by much.

32 'Those who work for Western aid and government agencies complain quietly of bugged telephones and fax machines, and office files which are rifled overnight.' *Far Eastern Economic Review*, 10 April 1994.

[33] According to *Asian Communications*, February 1994, a report by Detecon, the consultancy arm of Deutsche Bundespost Telekom, working on Vietnam's Telecoms Master Plan, with funding from the ITU and UNDP, recommended adherence to just two network hierarchies, trunk/international and local, controls over equipment types and usage rather than distance-based charging for data communications.

[34] HCM City, of which Saigon is the central commercial district, has a population of about 4 million, and Hanoi just over 1 million. Both cities also have a Lotus InterSputnik earth station.

[35] China seems to be following a similar course.

[36] 'Outward telephone connections will at the end of January quadruple to 100 international lines.' *Eastern Express*, 25 March 1994, Hong Kong.

[37] The *South China Morning Post*, 27 September 1994, quotes the fortnightly *Phnom Penh Post* as reporting that 'equipment to record phone conversations and fax transmissions was available in many local electronics shops and was widely used by the Government in an "aggressive crack-down on Government critics in general."' The Interior Ministry employed 800 staff for this task.

4 Companies and Capital in Asia-Pacific Telecommunications

Andrew Harrington*

The telecommunications industry throughout the Asia-Pacific region is undergoing major reforms which are bringing about changes to the structure and organization of the industry, including shifts of ownership and control from the public to the private sector, and opening the sector to domestic and international capital markets. In 1990, most of the region's telecommunications network and service industries were dominated by state owned or controlled monopolies and the number of publicly quoted telecommunications service companies was small. By 1993, many governments in the region had initiated structural reform including the introduction of competition and the encouragement of new operators, both foreign and domestic. Concurrent with these policy changes, there has been a significant increase in the number of telecommunications companies quoted on regional stockmarkets; by mid-1994 there were in excess of 25 quoted companies in the region (see Table 4.1). Furthermore, as Table 4.5 shows, many international telecommunications companies already have, or are actively seeking, equity participation in telecommunications development in the region. The purpose of this chapter is to put these developments into a perspective which examines the sources of capital finance required to promote the growth of telecommunications networks and services in the Asia-Pacific region, and assess how future changes are likely to influence the flow of capital into this sector.

---

* I am indebted to John Ure for his helpful suggestions on an earlier draft of this chapter. Of course, responsibility for the views expressed is mine alone.

Table 4.1
Quoted Telecommunications Service Companies in the Asia-Pacific Region, August 1994

Company	Country	Year Quoted
Hongkong Telecom	Hong Kong	1988
Champion Technology	Hong Kong	1992
Star Paging	Hong Kong	1991
ABC Communications	Hong Kong	1991
Philippine Long Distance Telephone	Philippines	NA
Philippine Telegraph and Telephone	Philippines	NA
Globe Telecom	Philippines	NA
Easycall	Philippines	1991
Time Engineering	Malaysia	NA
Technology Resources Industries	Malaysia	NA
Telecom New Zealand	New Zealand	1991
Telekom Malaysia	Malaysia	1991
Singapore Telecom	Singapore	1993
TelecomAsia	Thailand	1993
Shinawatra	Thailand	1991
Advance Info Services	Thailand	1991
United Communications	Thailand	1993
Loxley	Thailand	1993
Thai Telephone and Telegraph	Thailand	1994
Jasmine	Thailand	1994
Samart	Thailand	1994
Shinawatra Satellite	Thailand	1994
Korea Telecom	Korea	1994
Korea Mobile Telecom	Korea	1992
DACOM	Korea	1992
Videsh Sanchar Nigam Limited	India	1992
Mahanagar Telephone Nigam Limited	India	1992

NA = Not Available
Source: Salomon Brothers.

Structural Changes

The change in the ownership profile of the industry has coincided with a change in the basic characteristics of the companies that are operating within the telecommunications industry. As governments have liberalized, several new types of company have emerged with operations ranging from mobile telephony to second networks and value-added services. In many cases these are local companies which have established themselves as dominant players in other, often unrelated businesses, such as agriculture (for example, Charoen Pokphand in Thailand) or building construction (for example, Hutchison in Hong Kong), and being cash-rich, and also having very good connections with the political establishment, they are seizing the opportunity to enter one of the most profitable and rapidly expanding businesses in the region. As a consequence of this new entry, and of the more aggressive marketing strategies which greater competition has brought about, the telecommunications industry in the region is starting to resemble, in its structure and characteristics, any other industry in any other part of the world.

It is shifting from a state-run or corporatize utility towards a multi-entity competitive market structure, a shift that started in the customer premises equipment markets, such as telephone handsets and fax machines, and has since gone furthest and fastest in the value-added network services markets, such as paging and mobile telephones. In the fixed-wire public switched telephone network (PSTN) the shift has been more gradual and cautious, sometimes to no more than duopoly model, sometimes to less than that. But just the threat of competition, before it is fully realized, is already changing the behaviour of the dominant incumbent companies,[1] and as a result telecommunications products and services increasingly are being treated less as rationed items from the state and more like market commodities that might be found in any large supermarket.

The emerging industrial structure can be broadly classified into three major sectors:
1. The PTOs, or public telecommunications operators, are large companies serving a broad cross section of the market and commanding a dominant market share. Examples are Hongkong Telecom, the Philippine Long Distance Telephone and Telekom Malaysia.
2. The second networks operators are smaller with low, but potentially large market share in a large sector of the market. Examples are DACOM (Korea), Globe Telecom (Philippines), TelecomAsia (Thailand).
3. The mobile and value-added service companies are niche operators obtaining premium returns on capital and, usually, with a higher market share than the PTO. Examples, which include revenue-sharing entry into overseas markets, are Piltel (Philippines), Korea Mobile Telecom (Korea), Shinawatra (Thailand, Cambodia, Laos), Hutchison Telecom (Hong Kong), Champion Technology (Hong Kong/China).

This full scale industry restructuring is occurring simultaneously in many countries in the region and exhibits many common characteristics, such as amendments to telecommunications laws to permit foreign direct investment into some or all sectors of the industry, the liberalization of data services and the use and reception of satellite transmission, the licensing of second and third network operators, or alternatively of network service providers, and so on. That these changes, or discussions about changes such as these, are taking place across countries throughout the region suggests that there are fundamental forces shaping it. An empirical assessment of the political events surrounding the recent changes in each country's telecommunications industry would suggest that the fundamental reason why change is so widespread is that governments across the region have realized that the traditional state-owned or state-regulated monopoly structure of the telecommunications industry in these countries cannot support existing or desired levels of economic growth. This is true for all countries in the region both developed and developing.[2] This new interest in reforming the telecommunications sector in turn is a reflection of the desire of these countries to extend the economic achievements of the previous decade.

A more detailed examination (see below) suggests that the changes can be viewed as a basic restructuring, which has proved to be necessary to attract the capital and expertise required to support the industry's development. In this context, industry development is used in its broadest sense. In developed countries, where lines are readily available, it means lower prices and more services; in developing countries, where waiting lists are usually high, it means line growth.

In developed countries, typical changes have been the introduction of competition and the privatization of the network. This has served to increase efficiency and hence reduce prices and enhance the number of services available. Some of the new capital required has come from foreign sources (for example, Cable and Wireless in Macau, Bell Atlantic and Ameritech in New Zealand) but the availability of foreign capital was not a prerequisite for effecting the industry restructuring. This is because, in a developed economy, adequate supplies of capital will usually be available and there is often considerable expertise already present within the country. A good example of this would be Britain. Here foreign capital was present in the form of portfolio investment in BT (formerly British Telecom), Cable and Wireless and Vodafone, but it was not large and was not critical to the development of the industry.

By contrast, this is usually not the case in the developing countries, which are the major consumers of capital in the telecommunications industry in the region. We define developing countries as the Asia-Pacific countries excluding Australia, New Zealand, Japan, Hong Kong and Singapore. In these countries, the main policy priority is to promote line growth.

Competition for Capital in Developing Countries

Table 4.2 gives the expected increase in lines across the region between 1993 and 2000. Based upon a continuation of current projections and published expansion plans, an increase of around 148 million access lines is planned across the region by 2000.

Table 4.2
Growth of Access Lines in South East/East Asian region, 1993-2000E

Country	1993		2000 Estimated	
	Lines (millions)	Lines per 100 population	Lines (millions)	Lines per 100 population
Australia	8.5	47	11.5	59
New Zealand	1.5	44	1.8	53
Australasia	**10**	**47**	**13.3**	**59**
Hong Kong	2.9	48	4.6	77
Indonesia	2.0	0.9	11	5
Malaysia	2.4	13	7	25
China	20	1.5	80	5
Philippines	1.0	1.5	4	5
Singapore	1.2	43	1.8	59
South Korea	18	37	40	60
Taiwan	7.3	37	13	60
Thailand	1.5	2.7	5.5	9
South East/East Asia	**57.3**	**3.4**	**166.9**	**9.8**
India	7.5	0.9	20	2
Pakistan	1.2	1.0	3.5	2
India/Pakistan	**8.7**	**0.9**	**23.5**	**2**
Japan	**59**	**48**	**78**	**63**
Others[a]	2	0.6	3.5	1
Total Asia/Pacific	**137**	**4.4**	**285.2**	**8.5**
Developing Countries[b]	**63.9**	**2.2**	**187.5**	**5.9**
Developed Countries[c]	**73.1**	**48**	**97.7**	**63**

[a] Defined as Bangladesh, Brunei, Cambodia, French Polynesia, Fiji, Guam, Kiribati, Laos, Macau, Micronesia, Myanmar, Nepal, New Caledonia, North Korea, Papua New Guinea, Solomon Islands, Sri Lanka, Tonga, Vanuatu, Vietnam, Western Samoa.
[b] Defined as Asia/Pacific excluding Australia, New Zealand, Japan, Hong Kong, Singapore.
[c] Defined as Australia, New Zealand, Japan, Hong Kong, Singapore.
Source: International Telecommunications Union, Salomon Brothers.

In the developing countries, total access line additions could be 125 million, equivalent to an annual line growth of around 17 per cent per annum, and equal to around four times the total number of lines installed in the 1982 to 1992 period. In 1992, net profits in the telecommunications industry in the developing countries in the Asia-Pacific region stood at around US$6.7 billion per annum (see Table 4.3) with perhaps US$10 billion per annum of cash flows available for reinvestment, enough to purchase and install around 7 million lines per annum, equivalent to a line growth of around 10 per cent per annum. Hence, there is a shortfall between the amount of line growth that existing cash flows can support and the amount of line growth desired by companies and governments equivalent to a cu-

Table 4.3
Asia-Pacific Telecommunications Industry-Telecommunications Revenues and Net Profits in the Developing Countries, 1992 (millions of US$)

Country	Revenues	Net Profits
Indonesia	$1464	$308
Malaysia	1339	361
China	3864	1411
Philippines	682	195
South Korea	6660	818
Taiwan	4059	1482
Thailand	1467	425
South East/East Asia	**19535**	**5000**
India	2329	837
Pakistan	669	543
India/Pakistan	**2998**	**1380**
Others[a]	**850**	**289**
Developing Countries	**23383**	**6669**

[a] Defined as Bangladesh, Brunei, Cambodia, French Polynesia, Fiji, Guam, Kiribati, Laos, Macau, Micronesia, Myanmar, Nepal, New Caledonia, North Korea, Papua New Guinea, Solomon Islands, Sri Lanka, Tonga, Vanuatu, Vietnam, Western Samoa.
Source: International Telecommunications Union.

mulative total of around 60 million access lines.[3] Hence the telecommunications industry in the developing countries of the Asia-Pacific region are currently in a period when they will have to rely heavily upon outside funding if they are to achieve their telecommunications expansion ambitions.

This could well understate the situation because much of the cash flow deriving from existing operations is in local currency whereas purchases of equipment, which constitute around half of the total cost of network build-out, are likely to require direct or indirect US$ or equivalent funding. For some of the developing countries, conversion of local currency to US$ may be difficult in the amounts required although in some countries this is mitigated by the very high net international accounting rate settlements that they receive; for example in the Philippines some 42 per cent of PLDT's revenues is in the form of US$-based international accounting rate settlements.[4] Hence much of the funding gap discussed above may be required to be supplied in non domestic currencies.

Sources of Capital

There are five main sources of capital which can be utilized. The first is loans from commercial banks, multilateral agencies and national agencies.

The telecommunications industry represents an attractive industry for suppliers of loan capital because it has large and predictable cash flows and its dynamics are easily understood. However, with significant capital requirements growing in other sectors of developing economies, notably energy and transportation, and with the increasing demand for capital coming from a growing number of newly emerging market economies, especially in Eastern Europe, but also in Africa, and from more developed economies which in 1994 seem poised to emerge from recession, the amount of loans available to finance telecommunications network development may in the future be restricted. Furthermore, their importance as a source of finance for developing countries around the world has fallen significantly since 1981 as the international capital markets have developed and assumed a greater importance. In 1981, they provided around 82 per cent of all the external finance required by developing countries. In 1993, that share had fallen to 28.3 per cent (see Table 4.4).

Table 4.4
Net External Financing for Developing Countries (billions of US$)

Source of Funding	1981		1993	
	Billions of US$	As a % of total	Billions of US$	As a % of total
Multilateral Agencies	10.7	10.4	10.9	6.0
National Governments and Agencies	13.7	13.4	12.6	7.0
Commercial Banks	60.2	58.7	23.9	13.2
Non Bank Private Creditors	9.3	9.1	64.9	35.9
TOTAL DEBT	**93.9**	**91.5**	**12.3**	**62.1**
Equity Investments	8.7	8.5	68.5	37.9
TOTAL FUNDING	**102.7**	**100.0**	**180.8**	**100.0**

Source: Institute of International Finance.

The second source of capital is supplier credits, traditionally the single most important source of capital for network expansion around the world, although the total amount available is restricted by lending limits from the national export-import agencies of the developed countries which usually guarantee the credits. Given the extent of network development likely in the future in emerging economies around the world, supplier credits may well become harder to get and the best terms may be reserved for those countries and companies with the least perceived risks.

Domestic corporate equity and debt is the third major source of finance. Substantial investments by domestically owned corporations usually occur shortly after a liberalization of the domestic telecommunications industry. Examples include Shinawatra (Thailand), TRI (Malaysia) and Globe (Philippines). These investments usually provide an essential basis for network development, but they also provide a very rapid return for their owners, boosting their personal wealth as share prices on local stock markets often reach giddy heights. Telecommunications shares are easily hiked by the announcement of, or even the mere rumour of, a company being granted a new concession, licence or franchise. Price to earnings ratios in Asia have often over-reached themselves. For example, Singapore Telecom shares reached a P:E ratio of 25:1 immediately following flotation. Despite the inevitable element of hype and speculation, similar investments are likely to play an important part in network development in the future.

The fourth source of finance is strategic equity. This is likely to be in the form of strategic stakes by overseas telecommunications companies, and is usually accompanied by an infusion of employees from the strategic partner in order to oversee network development. By 1994 there was around US$3 billion invested in telecommunications projects in developing countries in the Asia-Pacific region by overseas telecommunications companies (see Table 4.5), mostly from developed countries where their networks are relatively mature and where they generate significant surplus cash flow. However, in the mid-1990s these companies are likely to be spending increasing amounts of money in their home markets on the development of new multimedia based services and in building a world presence in order to manage and support the global networks of their multinational clients. These projects can be expected to consume significant quantities of capital for the foreseeable future, so the number of companies with large surplus cash flows and the aspiration to invest them in developing country networks is likely to be relatively small, perhaps no more than 12 or so worldwide (see Table 4.6), with a maximum cumulative investment potential of, at most, US$5 billion by 2000.

The fifth major source of finance is portfolio investment. This consists of equity investments in quoted companies by institutional investors worldwide. In 1990, around US$1 billion was invested in telecommunications companies in the region, but by 1993 it had reached around US$15 billion and constituted perhaps 10 per cent of total institutional investment in the developing country stock markets. This source of funds is potentially among the largest of the five discussed, and is likely to play an extremely important part in financing network development. The telecommunications industry is intrinsically an attractive industry for investment because it is easily understood, and is a good barometer of the development of the broader economy, but enjoys relatively low risk. Portfolio investment is usually harnessed via an IPO or Initial Public Offering, the process by which a company achieves a quotation on a stock market. It is usually only available when most of the other four sources of capital discussed above have already been harnessed.

Table 4.5
Major-Cross Border Equity Investment in the Asia-Pacific Telecommunications Industry, 1990-94

Investing Company	Investment	Country	Comments
Nynex	TelecomAsia	Thailand	13.5% stake. 2 million line concession in Bangkok
NTT	TT&T	Thailand	18% stake. 1 million line concession in provinces
Singapore Telecom	Shinawatra Paging	Thailand	20% stake
Matrix	Easycall Thailand	Thailand	49% stake. Paging company
AT&T/McCaw	Smartone	Hong Kong	30% stake. Cellular network
Cable and Wireless	Hongkong Telecom	Hong Kong	58% stake. International and domestic fixed wire monopoly
Vodafone	Pacific Link	Hong Kong	30% stake. Cellular network
Cable and Wireless	Paktel	Pakistan	81% stake. Cellular network
Telstra	ICC	Philippines	40% stake. In early stages of building international gateway and 300,000 fixed wire lines
Singapore Telecom	Globe Telecom	Philippines	38% stake. In early stages of building international gateway, cellular network and 700,000 fixed wire lines.
Cable and Wireless	Eastern Telecoms	Philippines	40% stake. International gateway operator. Will build out 300,000 fixed wire lines.
Shinawatra	Isla	Philippines	30% stake. In early stages of building international gateway, cellular network and 700,000 fixed wire lines
Korea Telecom	TSI/Capwire/PT&T	Philippines	20% stake. In early stages of building international gateway and 300,000 fixed wire lines.
Millicom International	Extelcom	Philippines	40% stake. Cellular network. Inaternational gateway application pending.
Matrix	Easycall Philippines	Philippines	20% direct stake. Option to buy a further 26%. Paging company
Telstra	BCC with the Directorate General of Posts and Telecoms (DGPT)	Vietnam	Telstra will invest A$250m over a 10 year period to expand modernise the telecommunications network in return for a share in international revenues.
Shinawatra	Joint venture with Laos Government	Laos	70% owned by Shinawatra. 15 year concession to operate domestic, international, cellular and paging.
Matrix	Easycall Malaysia	Malaysia	49% stake. Paging company
Matrix	Easycall Indonesia	Indonesia	48% stake. Paging company
Airtouch	Shinsegi Mobile	Korea	10% stake. Second cellular network
Southwestern Bell	Shinsegi Mobile	Korea	7% stake. Second cellular network

Source: Salomon Brothers.

Table 4.6
Actual and Potential Strategic Investors in Telecommunications Projects in Developing Countries in the Asia-Pacific Region

Company	Country	Presence in Asia-Pacific?
Ameritech	USA	No
Bell Atlantic	USA	Yes. Regional office in Singapore. Owns 25% of Telecom New Zealand.
Bell South	USA	Yes. Regional office in Singapore. Joint venture with Jitong, a value added services network in China. Cellular network in New Zealand. 20% stake in Optus in Australia.
Bell Canada	Canada	Yes. Regional office in Hong Kong.
Pacific Telesis	USA	No
Nynex	USA	Ye. Regional office in Hong Kong. Owns 13.5% of TelecomAsia.
Southwestern Bell	USA	No
US West	USA	Yes. Regional office in Hong Kong.
Telstra	Australia	Yes. Has BCC in Vietnam and interests in Hong Kong and the Philippines.
Telecom New Zealand	New Zealand	Yes. Has stake in paging company in India.
NTT	Japan	Yes. Has stake in TT&T in Thailand.
Cable and Wireless /Hongkong Telecom	UK/Hong Kong	Yes. Has interests in the Philippines, Pakistan, Solomon Islands, Fiji and Vanuatu. Owns 20% of Optus in Australia. Owns 51% of CTM (Macau).
Singapore Telecom	Singapore	Yes. Has interests in Thailand and the Philippines.
Korea Telecom	Korea	Yes. Has interests in the Philippines.

Source: Salomon Brothers.

However, with the increasing range of investment choices now available, the availability of portfolio investment for a particular telecommunications project will increasingly depend on its attractiveness relative to other available, and competing, investments. With the global demand for capital enhanced by resurgent economic growth in the developed world, coupled with similar requirements for capital from developing countries in South America, Eastern Europe and, increasingly, Africa, and from other sectors within developing economies, capital is likely to be limited for telecommunications network development in the developing countries of the Asia-Pacific region. It is the extent to which capital from the sources described above is available that is likely to determine the shape of the telecommunications industry in the region.

Under these circumstances, a competition for capital may develop whereby reform programmes undertaken by countries in order to attract capital will be required to become ever more extensive to provide the conditions under which banks and multilateral agencies are prepared to lend, and corporate and portfolio investors are prepared to invest. These conditions would include a firm commitment by government to liberalize the telecommunications industry and, in particular, to encourage the participation of the private sector, both foreign and domestic. Ideally, also a stable regulatory environment with well defined terms of interconnection between carriers and a defined mechanism for regulating tariffs. An additional attraction would be the active involvement of a strategic partner, able and willing to commit capital and expertise to the project under consideration. Another condition is a low-risk country assessment, which means a favourable economic and political backdrop, including high economic growth, free market based economic policies and stable government.[5]

If these conditions are met, the evidence is that the project could provide the right combination of risks and rewards to attract the capital necessary for network development. It is possible that one source of capital may only become available if other sources of capital are already committed. For example, loan capital is likely to migrate to those countries and companies with the least perceived risks — a company in a developing country without a strategic stake by a foreign telecommunications company is likely to have less loan capital available to it than a similar company which does have a strategic stake.

The acceleration of reform within the industry, together with the emergence of a variety of different and innovative structures for private sector participation, provides powerful evidence that the competition for capital has begun in earnest. This competition is likely to intensify as more governments embark upon telecommunications reform, driven by knowledge of the success of reform programmes elsewhere and the inability of their existing network to support existing or desired levels of economic growth. The following is an assessment of the state of reform in the Asia-Pacific region.

Telecommunications Reform in the Asia-Pacific Region 1981-1993

In order to assess the trends within the industry, and their likely future, it is important to understand the changes that have taken place over the recent past. It is the purpose of this section to review these events. (See also Chapters 2 and 3.)

Throughout much of the 1980s the telecommunications industry in the Asia-Pacific region, with the exception of South Korea, showed steady but slow growth (see Table 4.7). Line growth averaged around 6 per cent, with the developing countries growing at around 16 per cent per annum and the developed countries at around 3 per cent.

Table 4.7
Asia-Pacific Telecommunications industry 1982–1991

Country	Lines (Mils)-1982	Lines (Mils)-1991	Lines per 100 population-1982	Lines per 100 population-1991
Australia	5.5	8.0	36	46
New Zealand	1.2	1.5	37	44
Australasia	**7.7**	**9.5**	**36**	**46**
Hong Kong	1.5	2.6	28	46
Indonesia	0.5	1.3	0.3	0.7
Malaysia	0.6	1.8	4.3	9.9
China	2.3	8.5	0.2	0.7
Philippines	0.5	0.6	1.0	1.0
Singapore	0.6	1.1	26	40
South Korea	4.1	15	10	34
Taiwan	NA	6.8	NA	35
Thailand	0.4	1.6	0.9	2.7
South East/East Asia	**10.5**	**39.3**	**0.7**	**2.5**
India	2.5	5.8	0.4	0.7
Pakistan	0.3	1.1	0.4	1.0
India/Pakistan	**2.8**	**6.9**	**0.4**	**0.7**
Japan	42.4	56.3	36	45
Others[a]	0.7	1.5	0.1	0.5
Total Asia/Pacific	**64.1**	**113.5**	**2.1**	**3.8**
Developing Countries[b]	**11.9**	**44**	**0.5**	**1.6**
Developed Countries[c]	**52.2**	**69.5**	**36.3**	**45.2**

[a] Defined as Bangladesh, Brunei, Cambodia, French Polynesia, Fiji, Guam, Kiribati, Laos, Macau, Micronesia, Myanmar, Nepal, New Caledonia, North Korea, Papua New Guinea, Solomon Islands, Sri Lanka, Tonga, Vanuatu, Vietnam, Western Samoa.
[b] Defined as Asia/Pacific excluding Australia, New Zealand, Japan, Hong Kong, Singapore.
[c] Defined as Australia, New Zealand, Japan, Hong Kong, Singapore.
Source: International Telecommunications Union, Salomon Brothers.

For as long as the telecommunications networks of the region were sufficient to sustain existing levels of economic activity, there was no imperative for governments to restructure the industry either to promote line growth in developing countries, or lower prices and more services in the developed countries. Reform had been attempted in some, such as the Philippines, where the monopoly of the Philippine Long Distance Telephone Company (PLDT) was formally abolished in 1986, but efforts were half hearted and they had little impact. South Korea was the exception to this. In

1982, the provision of basic services and data services was separated and two incorporated companies, 100 per cent state owned, were created, and major expansion programme was initiated so that by 1990, the number of lines in the country had increased 3.6-fold to 15 million (Table 4.7). Hong Kong liberalized its mobile services in the mid 1980s but a wider liberalization was restricted by the monopoly franchises for local PSTN and international network and voice service granted to Hong Kong Telephone and Cable and Wireless (Hong Kong) respectively in 1981. These companies merged in 1988 to form Hongkong Telecom.

During the latter part of the 1980s, two countries — Thailand and Malaysia — became concerned about bottlenecks in their telecommunications infrastructure and the drag effects this would have upon their economic and industrial growth. They embarked upon ambitious reforms, and although their reform programmes were different as a result of the different institutional and legal frameworks existing in the two countries, the net effect of both was a significant infusion of capital (see below) both domestic and foreign, into the telecommunications industry. The overall move towards reform was probably stimulated by the growing telecommunications reform movement in South America and the publicity that they received.[6]

In Thailand, the existing telecommunications legislation, principally the 1954 Telephone Organization of Thailand (TOT) Act and the 1976 Communications Authority of Thailand (CAT) Act, grants exclusive ownership rights for the entire telecommunications network to the state, through the TOT in domestic services and the CAT in international services (Table 4.8). As a

Table 4.8
Thai Telecommunications Legislation

Law	Year Passed	Details
Telegraph and Telephone Act	1934	Empowers the state to monopolize the provision of telegraph and telephone services to the public.
Telephone Organization of Thailand Act	1954	Established the Telephone Organisation of Thailand (TOT) as a state-owned enterprise to own and operate telephone services for the benefit of the state and the public. TOT subsequently acquired all the assets of the telephone division of the Posts and Telegraph Department (PTD) of the government.
Radiocommunications Act	1955	Authorizes the PTD to manage the use of radio spectrum. The PTD is formally responsible for the issuance of licences for all radio-based telecommunications services.
Communications Authority of Thailand Act	1976	Established the Communications Authority of Thailand (CAT) and transferred the monopoly on domestic non-telephone telecommunications services and international telephone services from the PTD to the CAT.

consequence, the traditional means of financing large scale infrastructure build-out, namely privatization and the introduction of competition, were not available.

The Thai government bypassed the legal restrictions by interpreting the legislation in a way that allowed the award of limited life concessions to the private sector. Under these concessions private companies build the telecommunications network, transfer ownership to either TOT or CAT and then operate it while giving a share of revenues and/or profits to TOT or CAT. By using this BTO or Build-Transfer-Operate approach, the state remains in formal ownership of all telecommunication facilities but private capital and expertise is attracted to the sector and network build out is promoted. Table 4.9 gives details of the main concessions granted under this procedure.

Table 4.9
Main Thai Telecommunications Concessions

Company	Sector	Concession obtained from	Major Shareholders	Comments
TelecomAsia	Fixed Wire-Bangkok	TOT	CP, Nynex	2 million lines. 25 year concession.
Thai Telephone and Telecommunications (TT&T)	Fixed Wire-Provinces	TOT	Loxley, NTT	1 million lines, 25 year concession.
Advance Info Services	Cellular	TOT	Shinawatra	20 year concession
Total Access Communications	Cellular	CAT	UCOM	15 year concession
Paclink	Paging	CAT	Pacific Telesis	10 year concession
Percom	Paging	CAT	Pacific Telesis	15 year concession
Easycall	Paging	CAT	Matrix Telecom	15 year concession
Phonelink	Paging	TOT	Shinawatra	15 year concession
Hutchison	Paging	TOT	Hutchison	15 year concession
Worldpage	Paging	TOT	Bangkok Electric Appliance	15 year concession
Fonepoint	CT2	TOT	Advance, UCOM	10 year concession

This approach has proved to be very successful with some US$10 billion of private sector capital (at market prices) having been invested or committed to the industry from both domestic and overseas entities and Thai telecommunications infrastructure improving dramatically, especially in the mobile telecommunications sector, where by mid-1994 there were 0.5 million cellular subscribers and 0.5 million paging subscribers. The improvements will be even more marked once TelecomAsia's 2 million line project in Bangkok and TT&T's 1 million line project in the provinces are complete. By then, the total number of access lines in Thailand will have almost tripled to around 4.5 million.

The Thai government has also taken steps to amend the 1934 Act abolishing the right of monopoly suggesting that, perhaps by the mid-1990s private sector operators should be able to build, operate and own telecommunications networks in Thailand. This will provide the basis for a further significant expansion in the telecommunications infrastructure and a further significant commitment of investment capital.

In Malaysia, reform initially began in 1984, when Telekom Malaysia was formed as a product of the government's privatisation policy. It was intended to be the holding company for the country's telecommunications assets which were formally transferred to Telekom in 1987. By the end of the 1980s, Telekom Malaysia's ability to continue its line expansion programme was being hindered by large debts from the past; relieved of this situation, its shares were floated on the Kuala Lumpur Stock Exchange in 1990. Around M$2.4 billion (US$960million) was raised for the company from the Initial Public Offering (IPO).

The regulatory environment under which the transfer of assets to Telekom Malaysia took place in 1987 did not specifically entrench a monopoly in any segment of the market. Telekom Malaysia's licence permits its operation in all sectors of the market in which it was present in 1987. These are fixed wire voice telephony, cellular, telex, data communications, payphones and equipment supply. In addition, regulations require the company to seek permission from the government in relation to any price increases and, in any event, restrict these increases to the domestic rate of inflation. There is also a universal service provision requiring the company to supply services throughout rural areas of the country.

Since 1987, the government has pursued a liberalization policy under which sectors of the market have progressively been opened up. However, the extent of liberalization, the terms under which new licences may be awarded, the degree of tariff flexibility allowed Telekom Malaysia, and the terms of interconnection between fixed wire carriers all remain unclear. Table 4.10 summarizes the mid-1994 industry situation.

Hence, by the beginning of 1993, the telecommunications industry in the Asia-Pacific region consisted of the following mix:
(1) two countries, Thailand and Malaysia, where reform was in progress;
(2) South Korea, where a government initiative had resulted in telecommunications development without significant liberalization;
(3) a number of territories, such as Hong Kong, the Philippines and Singapore, where reform had been limited, unsuccessful or non-existent.

Table 4.10
Malaysia's Telecommunications Market

Sector	Participants	Comments
Cellular	Telekom, TRI, Mobikom, Binariang, Sapura, MRCB	TRI is the market leader with 400,000 subscribers. Binariang has a GSM licence. Sapura and MRCB have both recently been awarded Personal Communications Network (PCN) licences.
Domestic Fixed Wire	Telekom, Time, Binariang, TRI	Time has already built out 1000 km of fibre optic cable along the North-South highway and will use this as a basis to construct a competing network. Binariang has not yet started to build it's network. It is unclear whether TRI has the right to offer competing fixed wire services.
International	Telekom, Binariang, TRI	TRI can use it's own gateway to carry calls initiated on it's cellular network but it is unclear whether they can compete directly with Telekom. Binariang can compete but is yet to build out it's network.

A Restructuring Emerges: Reforms From 1993

From 1993 many countries in the region significantly accelerated the restructuring of their telecommunications industry, driven by a number of factors which have came together. First, there was increasing evidence of the reforms that had occurred some years earlier in other countries, both within and outside of the Asia-Pacific region, were highly successful.[7] For developing countries, such as those in South America, this was measured in terms of the inflow of foreign investment capital to that region's telecommunications industry and the consequent degree of line growth that had been stimulated. In the developed world, such as Australia and New Zealand, this success was measured by the degree of competition that had emerged and the consequent reduction in prices and increase in the range of services on offer.

Second, developing countries in the region came to realize that, if they were to finance their ambitious telecommunications development projects, it was necessary for them to attract foreign capital and expertise. Furthermore, with a growing competition for capital, human resources and technology transfer from other countries, it was imperative that action be taken quickly so as to provide the investment returns necessary to attract overseas investment. Third, with the collapse of communism in Eastern Europe and the move towards market socialism in China, the greater use of market solutions became politically acceptable in many countries. As a result, virtually every country in the region has embarked upon, or has accelerated a reform-based restructuring of the telecommunications sector. Examples below illustrate the range of reforms that have been initiated across the region.

Hong Kong: Liberalization

In Hong Kong, the government capitalized on the opportunity presented by the end of the Hong Kong Telephone Company's (part of Hongkong Telecom) domestic PSTN monopoly franchise in 1995 to award three new licences to potential competitors in the local loop, which have been approved by the Anglo-Chinese Joint Liaison Group. The new licence holders are summarized along with their expansion plans in Table 4.11.

Table 4.11
Successful Bidders for Hong Kong Domestic Telecommunications Services Licence

Company	Shareholders	Amount to be spent	Comments
Hutchison Communications Ltd	80% Hutchison, 20% Telstra	HK$3.5 billion	Aims to build a fibre optic network to provide a range of advanced services to business customers
New T&T Hong Kong Ltd	Wharf Telecom	HK$6 billion	Aims to build a full service broadband network on the back of the HK$5 billion cable TV network already under construction.
New World Telephone	66.5% New World, 3.5% Infa, 25% US West, 5% Shanghai Long Distance Telephone	HK$2 billion	Aims to build a fibre optic network to support interactive voice, data and video. May combine it with a cable TV network

Source: Salomon Brothers.

It can be seen that all the new licence holders plan to offer a new range of services based upon broadband fibre optic technology and none plan to compete head on with Hongkong Telecom for basic voice services. This must be in part as a result of the interconnection arrangements between the new operators and Hongkong Telecom for the carriage of international calls. This was first outlined in October 1992 and provides for new operators keeping only 8.19 per cent of any outgoing international call revenues generated from their network. Since Hongkong Telecom's local loop charges are not sufficient to cover the cost of providing and maintaining the local network, a cross subsidy revenue from international call revenues is essential to the company's profitability. However, an 8.19 per cent revenue share for the competitors is likely to be insufficient to allow them to operate a profitable local loop network based upon basic voice telephony alone. Possible new offerings include video telephony, advanced data communications and interactive network services.

Hongkong Telecom International (another subsidiary of Hongkong Telecom) has a monopoly franchise for the international network and basic voice services until 2006. However, Alex Arena, the Director-General of the Office of Telecommunications Authority (OFTA) which was established in 1993 as the regulator, has clearly stated the intention of interpreting the licence as liberally as is legally possible to promote competitive international services. In line with this policy of opening up areas of the market not explicitly included in HKTI's licence, another development has been the awarding of a Public Non-Exclusive Telecom Services or PNETS Licence to BT to manage data networks in line with the company's strategy to offer international managed-data network services.

In the cellular market, a new network, SmartTone, was licensed to offer a digital cellular service based on GSM technology in January 1993. SmartTone is owned by ABC Communications, Sun Hung Kai Properties, McCaw Cellular and Town Kahn (a company owned by China's Ministry of Posts and Telecommunications). There are now four competing operators in the cellular market, with the promise of multiple entrants in the near future to operate personal communications networks using a variety of technologies. There are also four licensed operators of CT2, although only three offer a service, and in the paging market over thirty companies compete for a share of Hong Kong's one million plus market. In sum, the international market is restricted until 2006 but becoming more competitive, while the domestic market has become highly competitive in all but the basic service which will open after 1995.

Indonesia: First Steps

By contrast, Indonesia has taken just the first steps towards a comprehensive restructuring of the telecommunications industry which is specifically designed to attract large scale private sector capital and expertise. Prior to 1993, Indonesia had relied upon a number of small scale private sector projects under the PBH (Pola Bagi Hasil) programme to promote private sector investment into the industry. These projects provide for revenue sharing arrangements between the private sector and PT Telkom, Indonesia's state-owned domestic PSTN monopoly. The private sector partner builds the network and then transfers ownership to PT Telkom. Under the first PBH scheme from 1989, nine projects were undertaken using a Telkom-weighted 7:3 per cent revenue sharing split with a total line installation of around 775,000 lines.

Reforms were stepped up in 1993. The national Palapa satellite system was sold to PT Satelindo, a private sector company, and steps were taken for an IPO for PT Indosat, the state-owned international telecommunications corporation. The shares are likely to be quoted on the New York Stock Exchange. Thirdly, using World Bank funding, the Indonesian government commissioned a report from an international consultancy on how to restructure the industry to achieve PT Telkom's plans to expand the number of access lines by 5 million to around 8 million by 1999. This project — the telecommunications part of the national five-year economic development plan — is known as Repelita VI and will cost around Rp15,000 billion (US$7.5 million).

The report recommends that PT Telkom seek funding from the international capital markets and simultaneously, the company should seek a major expansion of private sector participation, including the involvement of foreign entities. Several schemes have been designed to encourage greater private sector participation. One such scheme is the Joint Operating Scheme or JOS, between Telkom and other companies, in which private sector companies invest in and operate key projects in return for a share of the revenues during the terms of the contracts. During Repelita VI, around 2 million of the 5 million lines to be built out will be allocated to the JOS scheme.

Malaysia: Regulatory Uncertainty

In Malaysia, the competitive environment changed significantly after 1993 as the government indicated that it wished to open up the part of the telecommunications industry that had hitherto been a *de facto* monopoly for Telekom Malaysia. This is, specifically, the fixed wire domestic and international voice sector of the market, currently worth some M$4 billion (US$1.6 billion). The precise number of licences awarded is unclear as is the interconnection regime with Telekom Malaysia and the degree of tariff flexibility that Telekom will have. A policy statement by the government in May 1994 raised more questions than it answered.

Numerous additional licences have been issued since 1993. A cellular licence was awarded to Mobikom, a joint venture between Telekom Malaysia (30 per cent), Sapura (10 per cent), PNB (30 per cent) and EON (30 per cent). This network, the third cellular network in , commenced operation in March 1994. Time Engineering, a company controlled by the Renong group, which is closely allied to the ruling party UMNO (United Malays National

Organization), was awarded a domestic fixed wire licence which enables the company to build a fibre optic network throughout peninsular Malaysia based on a fibre optic link that the company had built along the North-South highway. The network will be used to compete with Telekom in the provision of domestic telecommunications services to third parties. Technology Resources Industries (TRI) was given a licence to operate an international gateway switch in order to carry international calls originating from its wholly owned cellular network, Celcom, but it remains unclear whether TRI can use the switch to carry calls from other services and hence whether or not the company is competing directly with Telekom for international traffic. Also, Binariang was awarded the country's first digital cellular franchise to build out a GSM network. The company has also been given an exclusive licence to launch and operate Measat (Malaysia's first telecommunications satellite) and a licence to operate a domestic and international telecommunications network.

With this liberalization the government's policy priority clearly is to attract private sector capital and expertise into the industry to promote its development. This development will occur via investment in new network capacity as well as the introduction of new services. However, the way it is occurring — by the encouragement of competition in highly lucrative areas of the market — is more akin to those methods used by developed countries such as Hong Kong or Australia, than those common in developing countries such as Thailand or the Philippines whose policy priority has been to promote industry development across the board. Furthermore, the extent to which the policy can attract capital and expertise will depend crucially on the operating conditions under which the companies operate. Much of this remains unclear.

China: Opening Up

In 1993–94, there have been major steps taken towards a reform of the telecommunications industry in China. From a position where the Ministry of Posts and Telecommunications (MPT) had an absolute monopoly over all aspects of telecommunications, there have been a number of significant developments. These include the splitting of the regulatory and operating arms of the MPT, and the establishment in 1993 and 1994 of Jitong and Liantong, two new networks outside of the MPT's direct control. They are both strongly backed by the reform minded Ministry of Electronic Industries (MEI), which pushed through their establishment against very strong MPT opposition. For an analysis of these reforms, see Ure (1994a).

Jitong is owned by a group of 26 state-owned institutions including the MEI, the city governments of Beijing, Tianjin and Shanghai, the provincial governments of Jiangsu and Guangdong and the China International Trust and Investment Corporation (CITIC). It is engaged in seven national Value Added Network Services (VANS) projects and manufacturing ventures known as the seven 'Golden Projects'. These include a bank credit card payment and authentication network ('Golden Card'), a national information network of stored and real-time data linking state economic departments ('Golden Bridge') and a network connecting all customs offices ('Golden Gate'). In January 1994, Jitong signed a joint venture agreement with Bell South to provide network planning, design and engineering services for the communications businesses of Jitong.

The proposal to set up Liantong (to build a second long-distance network which will provide domestic long distance calls) was formally approved in March 1994. Liantong is controlled by the MEI, the Ministry of Railways, and the Ministry of Electric Power all of which operate their own private telecommunication networks. Other shareholders include CITIC, China Resources, China Merchants Holdings and the China Everbright International Trust and Investment Corp (see *Intermedia* February/March 1995, pp.22–24). Liantong plans to invest 100 billion yuan (US$12 billion) over the next five years to improve the capacity of the major shareholders own internal telephone networks. (See *Asia-Pacific Telecommunications*, August 1994.) It also intended to establish interconnection between these networks and the MPT's network so as to substantially increase the capacity of the public switched telecommunications network. Liantong intends that it should carry 10 per cent of the China's long distance calls and provide 30 per cent of its mobile telecommunications capacity by 2000. The MEI has recently signed a memorandum of understanding with Nynex which provides for the latter assisting Liantong in the design, engineering and planning of their network.

Other important developments include Order 128 of the State Council and Central Military Commission issued in September 1993. Order 128 maintains the separation of the State Radio Regulation Commission, which has overall responsibility for the allocation of public mobile radio frequencies, from the MPT, as well as confirming the control of the People's Liberation Army (PLA) over military spectrum, which in the future could easily be freed up for non-military uses. Related to this is the rapid growth of revenue sharing arrangements between PLA-owned paging and cellular networks and private and foreign telecommunications companies. And echoing the liberalization movement have been reports of important officials, such as the Mayor of Shanghai, Huang Ju (in 1994 to the Politburo of the CCP) at a conference in New York in November 1993, hinting that he favoured setting up a pilot project in which a foreign operator would take over operation of part of Shanghai's network.[8] It seems that the reform process has widespread support among the bureaucracy in the China and that the MPT is currently the only significant obstacle to its continuance.

Philippines: Reforms Promote Investment

The Philippines represents perhaps the best example of how reform became the only alternative available to a government anxious to promote telecommunications development as a means of promoting economic development. Despite industry liberalization in 1986, which removed the monopoly of the Philippine Long-Distance Telephone Company (PLDT), the telecommunications industry in the country remains dominated by Philippine Long-Distance Telephone (PLDT), which controls some 94 per cent of the country's telephone lines. Late in 1992, the government, led by President Ramos, launched a campaign criticizing PLDT for the lack of line growth and the long backlog for lines which amounted to around 800,000, or close to 100 per cent of the installed base. This campaign culminated in the government gaining a majority of seats on the board at the company's annual general meeting in April 1993.

Since that time the government has pursued a major programme of reform, central to which has been Executive Order EO59, enacted on 24 February 1993, requiring all authorized carriers to interconnect with each other and mandating that all subscribers should have the freedom to choose their long distance carrier. A further Presidential Executive Order, EO109, enacted on 12 July 1993, promoted universal access to basic telecommunications services by requiring providers of international or cellular services each to install 300,000 and 400,000 lines respectively over the next three to five years. These decisive measures have stung the PLDT to announce a 'Zero Backlog Programme', under which the backlog for lines will be eliminated by the end of 1996.

The government has indicated that the franchises of existing franchise holders will be interpreted liberally by the National Telecommunications Commission (NTC), which is the main regulatory authority in the Philippines. For new mobile, international and fixed wire networks, Provisional Authority (PA) to proceed with a business plan, and a Certificate of Public Convenience and Necessity (CPCN) to begin operations may be given by the NTC subject to the terms of the existing franchise. A number of powerful consortia have emerged as potential investors in Philippine telecommunications as a result of this initiative by the government (Table 4.12).

As can be seen, the plans are very ambitious and, if they are all completed, are likely to result in the infusion of at least US$5 billion into the Philippine telecommunications industry by 2000.

Singapore: An IPO

In Singapore, the Singapore Telecom IPO took place in October 1993, completing a process initiated in the late 1980s, when the government announced

Table 4.12
The Philippine Telecommunications Market — New Participants

Consortium	Local Participants	Foreign Participants	Awards	Comments
Globe	Ayala Corporation	Singapore Telecom (38%)	Cellular, International Gateway	Required to build out a minimum of 700,000 fixed-wire lines. Allocated areas include Metro Manila's financial and business district.
Smart	MetroFacific Corporation		Cellular, International Gateway	Required to build out a minimum of 700,000 fixed-wire lines. Allocated areas include Manila's outlying residential and industrial districts as well as the special economic zones of Subic and Clark.
Islacom	Citadel Holdings/ Delgado Family	Shinawatra Group (30%)	Cellular, International Gateway	Required to build out a minimum of 700,000 fixed-wire lines. Allocated areas include the whole of the Visayas.
ICC	Benpres Holdings/ Lopez Family	Telstra (40%)	International Gateway	Required to build out a minimum of 300,000 fixed-wire lines. Allocated area includes Quezon City.
Eastern Telecommunications	Government (original stake sequestrated)	Cable and Wireless (40%)	International Gateway	Required to build out a minimum of 300,000 fixed-wire lines. Allocated area includes the western part of Manila and North East Luzon.
TSI/Capwire/PT&T	Santiago Family	Korea Telecom (20%)	International Gateway	Required to build out a minimum of 300,000 fixed-wire lines. Allocated area includes Quezon and Laguna.
Piltel/Philcom	PLDT, JAKA Group	Comsat (16.3% of Philcom)	International Gateway	Required to build out a minimum of 300,000 fixed-wire lines. Allocated area includes Mindanao.
Digitel	JG Summit		International Gateway Application Pending	Discussions are taking place with Telecom New Zealand. Already has national fixed-wire franchise.
Beltel	Ortigas Family		International Gateway Application Pending	Discussions are taking place with Bell South.
Extelcom	Philcomsat	Millicom (40%)	International Gateway Application Pending	Already has nationwide cellular franchise.

Source: Salomon Brothers.

its intention to privatize the company. Around 10 per cent of the company was sold, with much of the equity going to domestic retail investors. The government also established Singapore Telecom's regulatory environment through the creation of the Telecommunications Authority of Singapore (TAS). The company will be a monopoly in mobile services until 1997 and in all fixed wire services until 2007.

One of the main aims of the privatization was to stimulate interest in the Singapore equity market by providing a large liquid stock, a good proxy for the state of the Singapore economy. Another aim was to provide a commercial environment for the company so that it may compete aggressively for hubbing business from international companies. To this end Singapore Telecom's tariff flexibility is determined with reference to a basket of published tariffs from other countries with whom Singapore Telecom is competing. These are principally Hong Kong, Australia and Japan. See Hukill (1994).

South Korea: Bold Steps

Despite an early start to reforms in 1981, until recently a paralysis of policy set in, but now this seems to be giving way to a new determination. The government has indicated the wish to promote a substantial liberalization by announcing its intention to abolish the current regulatory division of the industry into General Service Providers (GSPs), which provide a broad array of telecommunications services and Special Service Providers (SSPs), which can provide only value added services. The GSPs have hitherto had shareownership and other restrictions that the SSPs have not had, and from 1995, SSPs and GSPs will be able to offer the same array of services. Currently the GSPs include Korea Telecom, while the SSPs include DACOM and Korea Mobile Telecom. The government has also indicated a wish to change some of the restrictions applicable to GSPs to enable foreign investors to own equity in them. Currently foreign equity ownership is not permitted in GSPs and is restricted to 10 per cent of SSPs. A change in this rule would enable an IPO of Korea Telecom.

Some bold policy steps have already been taken, such as the sale of 23 per cent of Korea Mobile Telecom (previously 65 per cent owned by Korea Telecom) to Yukong Limited, a member of the Sungkyong Group. Also 15 per cent of Korea Telecom was sold to non-institutional Korean investors in late 1993 and early 1994. And despite a scandal surrounding the award of the second cellular network licence to a consortium led by Sungkyong in 1992, in which Presidential family ties seemed to be involved, the ownership of the second cellular network was resolved in 1994 with the award going to another consortium of Korean and overseas business interests including POSCO, Kolon, Airtouch and Southwestern Bell.

Thailand: Fruits of Success

As discussed earlier, Thailand was among the first countries in the region to embark upon a major reform of its telecommunications industry, and 1993 was the year when these reforms started to bear fruit. In the PSTN, TelecomAsia's 2 million line concession in Bangkok was successfully financed on a project basis by a mixture of debt provided by equipment suppliers and commercial banks and equity provided by the CP (Charoen Pokphand) Group and Nynex. In addition, the company successfully raised an additional Bt12.3 billion (US$490 million) from international investors in their IPO for projects outside the basic concession. Also in PSTN, Thai Telephone and Telecommunications (TT&T) raised Bt13.8 billion (US$550 million) of debt from equipment suppliers and commercial banks in order to finance their one million line concession in the provinces outside of Bangkok. The rest of the financing came from equity sources including some Bt7.3 billion (US$290 million) from institutional investors during their IPO.

A number of other telecommunications companies also had successful IPOs in 1993 and 1994 including Shinawatra Satellite, United Communications, Loxley, Samart and Jasmine International. These value-added service companies raised around Bt5 billion (US$200 million) from domestic and international investors. The number of cellular subscribers doubled from around 250,000 at the end of 1992 to 0.5m subscribers by mid-1994 and cellular penetration is close to 1 per cent of the population. But the biggest reform is yet to come, as the government has indicated a desire to privatize both TOT and CAT and has appointed consultants to advise them on the most appropriate way to proceed.

Vietnam: Cautious Approach

In Vietnam, the first concrete signs of reform have appeared in the form of the Business Cooperation Contract (BCC), the only form of foreign participation currently permitted. The Australian PTO, Telstra, has committed to a A$250 million (US$185 million) investment under a BCC over ten years in the expansion and modernization of Vietnam's telecommunications network. It includes international gateway and domestic exchanges in Hanoi and Ho Chi Minh City, provincial exchanges, transmission equipment, fibre optic submarine cable system and a programme of training and management assistance. In return, Telstra receives a share of the DGPT's (Directorate-General of Posts and Telecommunications) revenue from international telecommunications services over the period. Telstra has been involved in Vietnam since 1986 with its first BCC signed in 1988.

A number of trials, which may lead to BCCs, are underway in radio-based value added services, such as paging, trunk radio and mobile cellular. Besides these, foreign investment in Vietnam's telecommunications sector is limited to equipment supply, such as TelecomAsia's memorandum of understanding, signed in late1993, with Ho Chi Minh City Post and Telecommunications Authority to install switching and other network expansion equipment to provide at least 100,000 lines in Ho Chi Minh City. Vietnam has been considering ways to amend and update its telecommunications regulations, including drafting a telecommunications law, but with a very underdeveloped economy only cautious progress is being made.

Corporate Structures in Asian Telecommunications

The restructuring of the Asia-Pacific telecommunications industry that has occured over the last few years, in response to the competition for capital discussed above, has resulted in a number of different corporate structures emerging within the industry. They are summarized in Table 4.13.

The range and diversity of these structures indicate the differing policy constraints that governments operate under. These constraints derive from the perceived strategic importance of the telecommunications industry and centre around two key areas: the degree of private, including foreign, ownership that is permitted, and the degree to which foreign involvement in management is permitted. These must be set against the desire of private, including foreign, investors to maximize investment returns while minimizing investment risks.

There are a number of countries which have no constraints on ownership or management. These include all the developed countries in the region as well as India and Malaysia. However, licences are usually required and these may place modest constraints on the operation of the networks and/or make provisions for the imposition of government restrictions in certain extreme circumstances. Other countries, including Thailand and Indonesia, have legal restrictions on ownership and management but have interpreted these rules in such a way as to make an investment in their telecommunications industry by the private sector an attractive proposition.Usually these constraints derive from historic legislation that is in the process of being revised.

In other cases, such as the Philippines and China, governments publicly seek to justify the constraints on the grounds of 'national security' but privately are prepared to reinterpret the rules in various degrees within the existing framework.Ultimately, the driving force behind the emergence of these structures and the desire of governments to be flexible within existing frameworks is the need to attract capital in order to invest in their telecom-

Table 4.13
Corporate Structures in Asia-Pacific Telecommunications

Structure	Examples	Description
Build Transfer Operate (BTO)	Thailand	Designed to accomodate Thai law which prohibits any private ownership of telecommunications assets. Ownership of the project assets is transferred to the TOT immediately after construction. the concessionaires then operate the assets in return for a revenue share.
Build Own Operate (BOO)	Hong Kong, India, Malaysia, Japan, Australia, New Zealand	This is the most common form of private sector participation, particularly in developed countries. Few or limited restrictions on ownership. They usually require a licence issued by government.
Business Cooperation Contract (BCC)	Vietnam	These are relatively tenuous agreements establishing an intention to cooperate in relation to a particular project. In Vietnam, BCCs in telecommunications have tended to involve a degree of revenue sharing between the foreign participant and the Vietnamese partner. In some ways, BCCs are similar to a Memorandum of Understanding.
Build Maintain Transfer (BMT)	Philippines	These are a form of supplier financed network build out under which the equipment supplier builds the network, finances it, maintains it and leases it to the operator until the end of the lease.
Joint Operating Scheme (JOS)	Indonesia	These are agreements whereby the private sector consortium builds and operates a network in return for a revenue share with the PTO. In Indonesia, the revenue share is 70%.
Build Transfer	China, Indonesia	These are common in countries where private sector participation in operating a network is forbidden. It is essentially an equipment supply contract although in many cases it also involves the equipment supplier receiving a share of network revenues instead of having a formal leasing agreement which is the more common method of receiving supplier finance.
Build Lease Transfer	China	Hitherto, this has been common within China whereby domestic corporations with spare funds purchase telecommunications equipment and lease it to the local Posts and Telecommunications Adminitrative Bureau in return for lease payments and revenue sharing prior to transfer. A number of foreign corporations, including Hongkong Telecom are currently in discussions regarding the setting up of similar arrangements where the payments accumulate and are convertible into equity at some future date.

Source: Salomon Brothers.

munications industries. The less restrictions on operations and ownership, the more capital will be available for investment.

Conclusion

Most of the countries in the region are now in the process of restructuring their telecommunications industry. For example, India is embarking upon a quantum leap in reform, opening the market to foreign direct investment, issuing new licences, and promoting competition. The scale of reforms on the sub-continent are bound to impact upon the thinking of Asian regulators and on the regional competition for capital.

Throughout the region, there is a wide variety of reform programmes in place, but they have certain features in common. First, a liberalization of the industry involving the opening up to competition sectors of the market that had hitherto been restricted. The most common areas for initial reform are usually mobile and value added services, although it can be seen that many countries now permit some form of competition in all areas of the market.

Second, an opening of the industry to sources of capital that had hitherto been untapped. In most cases, these new sources consist of foreign corporate or institutional capital and/or domestic institutional capital. It is usually effected by an Initial Public Offering of shares of incumbent operators on the local and international capital markets.

Third, a willingness to create new forms of corporate structures in order to promote network build out in sectors of the market that otherwise may not be built. These include Build-Transfer-Operate (BTO), Build-Operate-Transfer (BOT), Build-Lease-Transfer (BLT), Build-Transfer-Manage (BTM), Business Cooperation Contract (BCC), revenue sharing arrangements, equipment leaseback agreements, licences, and so on.

The driving force behind this restructuring has been the desire to attract capital from overseas sources for investment in the telecommunications industry in the region. This has derived from the very substantial telecommunications expansion programmes of the developing countries in the region in the period up to 2000, which are significantly in excess of their expansion in the 1980s, and their inability to fund three programmes from internal resources.[9]

Competition for capital has developed and resulted in an industry restructuring. In 1993, the restructuring process accelerated and, as a consequence, during 1994 there was a significant reform of the industry in a number of countries, opening the door to corporate investment and a rush of issuance in the equity market. This acceleration will continue, resulting in substantial further opportunities for providers of debt and equity capital around the world as well as for private and strategic investors. During this time, countries will be driven to compete with each other for investment by providing ever more favourable regulatory and operating environments, which reduce investment risks and increase investment returns. Furthermore, in order for the capital required to be successfully raised it is likely that many of the legal constraints that have held back the degree to which countries have been able to reform will disappear within the next few years.

---

1. In economic literature, the theory of contestable markets has been used to argue that just the threat of competition is sufficient to force dominant companies to adjust their behaviour and act as if they were operating in a competitive market. Of course, for the threat to be real, either entry has to be allowed, or the state has to establish trigger points which will open the market if the dominant player fails to meet pre-set targets. See Baumol, Panzar and Willig (1988).
2. Even Singapore, where Singapore Telecom provides a world-class service so far as technology is concerned, remains the target of multinational companies who complain about bureaucratic inertia. See Cureton (1992).
3. A useful rule-of-thumb is to assume a cost of around US$1500 per line to purchase and install an average line, a typical amount for a developing country with low line penetration. This translates into a cumulative funding gap of up to US$90 billion by 2000. This is in contrast to the 1982–92 period when cumulative cash flows from operations were sufficient to finance a total telecommunications investment of around US$60 billion throughout the period. All statistics are derived from ITU (1994).
4. By convention, international carriers pay each other according to the amount of traffic the transmitting carrier sends for termination, or for transit through the receiving carrier's territory. An accounting rate negotiated between two carriers establishes the per minute transit or termination charge, for example US$1 per minute of traffic, while the settlement rate determines what proportion of the accounting rate is to be paid. Usually, but not always, this is 50 per cent, for example 50 cents per minute paid by the sending carrier to the receiving carrier. When outgoing and incoming traffic streams on a particular route balance each other, net payment is zero. In poor countries international call charges are high so users have an incentive to arrange for calls to be incoming, and this results in high net inward accounting rate payments to countries such as the Philippines.
5. Economic and political risk varies widely across Asia. Frey (1984) summarizes evidence which suggests that per capita income and balance of payments figures are strong indicators of economic stability, while loans from commercial and multilateral agencies such as the IMF and the World Bank are generally good political risk indicators, irrespective of whether the regime is of the left or right.

6   Petrazzini (1993) compares the process of telecommunications reform in the Argentine and Mexico with reform in Indonesia and Malaysia, suggesting a relationship between the degree of central state authority and the success of reform. In a critique,Ure (1993) places greater stress on the development issue.

7   Besides numerous articles in specialist telecommunications journals and the financial press reviewing the success of liberalization, significant assessments have been published by multilateral lending agencies. For example, the IFC (1990), and the World Bank (1989; 1992a; 1992b; 1994a, 1994b) and the WB/IFC (1994).

8   Since late 1992, China has allowed Build-Transfer-Operate (BTO) schemes involving foreign direct investment in pilot infrastructural projects, but not yet in telecommunications. Shanghai's mayor, Huang Ju, was reported as saying 'We would like to designate a place to experiment with more international participation in the telephone network.' (*South China Morning Post*, 12 November 1993, p.B2.) Shanghai, like the rest of the China, has ambitious telecommunications expansion goals. By 2000, they wish to quadruple their network from 1.5 million lines to around 6 million lines.

9   However, some authors have suggested that the pool of domestic savings may, in some circumstances, help alleviate the need to seek overseas funding. For example, in the case of China, Ure (1994b) argues that domestic revenues and foreign non-direct investment may meet financing requirements.

Information Highways and the Trade in Telecommunications Services

Peter Lovelock*

Until quite recently, the common wisdom was that, unlike most goods, services could not simply be shipped from one market to another. Services were intangibles (in terms of account ledgers, they were often also 'invisibles') that were produced and consumed simultaneously, usually in the same place. If an insurance agency, a financial house or a television broadcaster, wanted to sell its services abroad it would have to move to its customers; it would have to establish a local operation in the foreign marketplace.

The evolution of information economies and transnational communications networks, however, has changed all this. (See Ducatel and Miles, 1992.) The location of a worker producing or providing a service, for example creating an advertisement or entering data, becomes substantially less important if a company and its customers are connected through a communications network. As a result, there has been a huge growth in the internationalization of service industries through the 1980s and 1990s. Many traded services such as advertising, accounting, insurance, and banking and

---

* The author would like to thank the following individuals for their comments, suggestions and critiques on earlier drafts of this chapter: Michelle Cutler, John Langdale, Ben Petrazzini, Susan Schoenfeld and John Ure. The usual waiver applies, in that responsibility for all opinions expressed remain the author's alone.

financial services, have become not only transnationally feasible, but highly profitable as the result of transnational communications networks. (See Cowhey and Aronson, 1993.) The examples are numerous: animation for the Japanese television industry is drawn offshore in southern China; abstracts for American legal databases are contributed by South Koreans; commercials conceived of in Sydney are adapted for use around the world.

However, even though a services firm may not need to set up a foreign subsidiary or operation, it will still need to reach its customers in the foreign country. This will, of course, be done via telecommunications networks. And it is this which has proved problematic in negotiating trade-in-services agreements for telecommunications: for not only is telecommunications a service sector in its own right, it is also a mode of delivery for all services. Indeed, the use of communications networks to move information underpins virtually all service transactions, and this is why the communications sector has figured so prominently in both regional and global economic arrangements, such as the GATT (General Agreement on Tariffs and Trade) negotiations.

Early debates in the GATT talks centred on the question of how to differentiate between basic services and enhanced (or value-added) services on the presumption that only the latter were traded. This distinction proved impossible to maintain. Rapidly changing technology and the unilateral decision by some countries to liberalize basic services, while others remained stuck on definitions of value-added, meant that discussions at the GATT shifted to focus on the modes of service delivery. The dominant form is cross-border traffic. See Table 5.1.

Table 5.1
Trade in Telecommunications Services

Mode of service delivery	Example for telecommunications	Significance for telecommunications
Cross-border	Basic international traffic for voice telephony, telex, X.25, etc. International Value-Added Network services.	Cross-border trade is dominant made of telecommunications traffic.
Establishment of business operations	Foreign-owned company offering national and international telecommunications service from within country.	Growing importance as market access and foreign ownership restrictions are lifted and infrastructure competition is permitted.
Movement of customers	Roaming agreements for mobile operations (eg, GSM). Calling card and city-direct services.	Growing importance, particularly if global mobile satellite services are established.
Movement of labour	Build-Operate-Transfer (BOT) schemes, training and other consultancy operations.	Minor importance except as an add-on service to telecommunications equipment contracts.

Source: ITU, World Telecommunications Development Report (Geneva, 1994)

By 1992, the trade in commercial services internationally was estimated to be worth US$915 billion, and to be growing at approximately 6 per cent per year.[1] In December 1993, the Uruguay Round of the GATT was eventually completed and, for the first time, a General Agreement on Trade in Services — GATS — was added. As with GATT and manufactured trade, GATS is essentially a code of orderly conduct for trade in some of the major service sectors of the global economy.

Telecommunications was left out of the General Agreement, but a 'sectoral annex' relating to enhanced services was included.[2] (See Apendix on p.243.) The annex targets the removal of national regulations such as the number of firms allowed in a market, economic 'need' tests and local incorporation rules which favour domestic as against foreign firms in enhanced telecommunications services and supplies. This is expected to have its greatest impact on countries with state-owned monopolies such as Singapore, Taiwan and South Korea.

In 1992, the International Telecommunication Union (ITU) estimated the size of the global telecommunications market to have reached US$535 billion. Of this, services accounted for more than three-quarters of the total market, some US$415 billion. Only one-quarter of the market, trade in telecommunications equipment, remains relevant to international service provision in three specific ways:

1. Unless telecoms equipment is compatible it may be impossible to exchange certain services internationally. (Hence the often aggressive and apparently loss-making strategies of vendors to be first into markets such as China.)
2. The ability to deliver services influences the sale of equipment. The attractiveness of AT&T's most sophisticated digital switches, for example, depends on countries permitting the creation of '800 number services' similar to those available in the US.
3. Some services will not work without specially designed equipment. If the PTT (Post, Telegraph and Telephone authority) forbids attachment of the equipment to the network, the service cannot be delivered.

Telecommunications Services

The definition of telecommunications services remains problematic as there are no clear lines separating communications, information services, broadcasting and even publishing. For example, if a digitalized telephone call travels as a string of data, is it data communications? Charges for a computer transmission linked to the public switched telephone network (PSTN) via a modem will be the same as those for a telephone call, but is it one? A satellite news programme is picked up and retransmitted to residential receivers over optical fibre cable lines, but is this a broadcast (media) service or telephony?

Telecommunications services are usually classified as either basic or enhanced. There are, as indicated above, problems with this distinction. Enhanced services are themselves broken down into value-added services and information services. Fundamental to the provision of telecommunications services are infrastructural *facilities*, which include the physical infrastructure for communications (the transmission line and switching) and the facilities *services* (transmission services rented, or leased, from someone with physical infrastructure). This distinction between facilities and services has received more and more attention in trade negotiations. With the introduction of broadband facilities, mobile roaming and private network services, it is a distinction that is also rapidly becoming redundant.

Telecommunications services include basic services (the transmission of a message whose content and format remain essentially unaltered, as in a phone call), value-added services (which alters the format of a message in order to make it more 'user-friendly', or more efficient) and information services (the processing of the content of a message).

In this chapter, each of these components will be addressed. By following the telecommunications traffic through the satellites, cables and, increasingly, mobile applications, the emergence of a regional communications infrastructure in East and Southeast Asia will be examined. An infrastructure, which, it is posited, will be as important to regional economic developments as were transportation networks previously. The trade and political relations between the ten neighbouring countries will be examined in this context. The focus here is therefore *regional*; unless related to specific themes, domestic considerations have been left for discussion elsewhere in this volume. The chapter is essentially divided into two parts: regional facilities and facilities services.

Facilities Services

International telecommunications traffic is the fastest growing part of the telecommunications business and also the most profitable. International telephone traffic has been growing at some 17 per cent per year over the last decade, fuelled by growth in international trade, travel and immigration and a general expansion of the global economy. In 1992, 42.5 billion minutes of international telephone traffic were recorded. (See IIC, 1993, p.32.) Significantly, the fastest growing part of the traffic is not voice, but data and image. The traffic patterns of Hong Kong are indicative: fax and data accounted for less than 2 per cent of outgoing telephone traffic in 1983; by 1992 this had risen to 15 per cent, an annual growth rate of 27 per cent.[3]

In terms of the impact this will have on the provision of services to, and within, Asia, the statistics are enlightening. As Table 5.2 shows, while Asian communities are communicating more internationally, they are — almost without exception — 'talking' for less time. The reason? Access to fax and telex services. According to the data collected by APEC the length of the average call/message fell by a standard 30 to 40 per cent between 1985 and 1992 for *all* countries under review.[4] (See APEC, 1994a.) This is just one example of the technological leap now being attempted by the fast-growing Asian economies — the tools of the 'information revolution' are rapidly becoming available across the Asian region.

Table 5.2
Average Length of Call/ Message

From To	Hong Kong 1985	Hong Kong 1992	Taiwan 1985	Taiwan 1992	Malaysia 1985	Malaysia 1992	USA 1985	USA 1992
Australia	5.8	4.1	4.9	2.6	6.9	4.6	9.5	5.7
Canada	7.3	5.7	–	3.4	8.8	4.8	–	4.9
China	5.0	3.2	–	3.1	8.0	2.8	11.7	7.8
Hong Kong			3.7	2.3	5.6	3.0	8.8	4.8
Indonesia	4.7	3.0	4.4	2.5	5.3	3.4	12.1	7.3
Japan	4.3	3.3	3.8	3.1	5.6	–	8.4	6.5
South Korea	4.5	2.4	3.9	2.2	6.2	3.1	9.0	7.1
Malaysia	5.4	3.2	4.5	2.7			12.7	7.3
New Zealand	5.0	3.7	5.2	2.6	8.6	4.5	10.9	5.6
Philippines	3.9	3.7	4.0	2.8	6.1	3.7	11.6	10.6
Singapore	4.7	3.2	4.1	2.7	–	–	8.7	4.5
Taiwan	3.8	2.5			6.1	2.9	8.5	5.6
Thailand	5.0	3.0	4.8	2.7	4.6	3.0	12.1	7.4
USA	6.9	3.8	5.5	3.8	8.7	4.1		

Source: APEC, 1994a.

But while the average length per communication has fallen, the absolute amount of international communication has grown dramatically (Table 5.3). One reason has been regional migration patterns from (and to) Asian countries. The new immigrants may have experienced a hostile reception in some countries, they have been a boon to telephone carriers, as Telstra discovered when they installed their first satellite earth station in Vietnam in 1986. Latent demand, the desire to 'phone home', finally found a means of expression.

The key drivers of international telephone *and message* traffic, other than migration flows, are price, economic activity (GDP), and trade. The close relationship between telecommunications and trade is an important

Table 5.3
International Telecommunications Traffic Growth in East Asia, 1992

Country	Growth (1991–92)	Country	Growth (1991–92)
China	44.3%	Singapore	19.8%
Hong Kong	24.9%	South Korea	27.5%
Indonesia	18.1%	Taiwan	31.9%
Malaysia	22.1%	Thailand	19.8%
Philippines	16.1%	Vietnam	NA

Source: IIC, *Telegeography 1993*.

point and one which is developed later in this chapter (Tables 5.4 to 5.7). As a regional economy continues to emerge in Asia, telecommunications traffic flows *within* the region can be expected to grow relative to traffic with the rest of the world. This will provide the justification (or economic incentive) for further facilities and greater provision of services.

An Emerging Regional Infrastructure: Trade and Traffic

Intra-Asian Trade

Even though recent Asian economic growth is often seen as being the result of labour-intensive exports to the West, intra-Asian trade grew faster through the 1980s than inter-regional trade with either North America or Europe (Figure 5.1). Average annual export growth from Asia to North America measured 10.8 per cent and to Europe 10.5 per cent, while intra-regional exports grew by 11.4 per cent. The result of this rapid growth in intra-regional trade is that Asia now accounts for a much greater share of total Asian exports than either the USA or Europe. Asia's share (not including Japan) increased steadily from 23.3 per cent in 1981 to 35.9 per cent in 1991. (If Japan is included, the proportion rose from 43.1 per cent to 49.5 per cent.)

A better indicator of emerging regional economic interaction may be foreign direct investment (FDI) flows. FDI has become the biggest single source of financing and, according to the World Bank, the most potent development tool in East Asia (World Bank, 1993). As opposed to the comparative, or complementary, advantage upon which the trade in goods (exports and imports) is based, FDI is now seen as indicating a longer term commitment to a host country's development. The World Bank believes foreign investment offers a much faster route to economic development

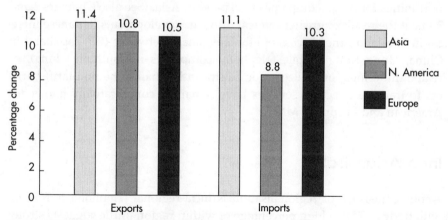

Source: EIU, *Business Asia*, 30 August 1993, p.4.

Figure 5.1 Annual percentage change of Asian imports and exports, 1981–91.

than policies which seek to promote and protect infant indigenous industries. This is largely because foreign investment in East Asia has been increasingly oriented to small and medium-sized enterprises, mainly of East Asian origin. As a result, investors are no longer seen as such a threat to national sovereignty.

In the period from 1986 to 1991, combined American and European investment accounted for approximately 20 per cent of total FDI in East Asia. (A large part of this took the form of reinvestment of profits.) More dramatic, however, was the activity of the newly industrializing economies in Asia — Hong Kong, Singapore, South Korea and Taiwan — which boosted investment to 46 per cent of the total. (Japan accounted for 22 per cent of the total.)

As a result of their rapid opening to global (and regional) economic forces, pressure has been strong from the business communities across Asia to gain access to telecommunication services at cost-based prices. The integration of the international and domestic economies has thus resulted in structural adjustment — political, as well as economic. Petrazzini, for example, has noted that, 'pressure for the liberalization of the market is paired with the general inability of LDC state-owned telecom companies to provide those services as rapidly as demand requires' (Petrazzini, 1993, p.7). This in turn has made governments across the region even more outward looking in their bid to sustain rates of economic growth. Multinational operations, particularly in the form of joint ventures or BOT (Build-Operate-Transfer) schemes, have become increasingly acceptable. (See also Chapter 4.) An interesting development in this regard is that, increasing intra-Asian trade has been reflected in a greater neighbourly interest in infrastructural developments, and this has often translated into greater opportunities for Asian companies to expand in Asia. (See also Chapters 2 and 3.) Be it the result of greater interest in Asian developments (Japanese interest in Thailand), the success of ethnic business networks (CP Popkhand in China, Hong Kong mobile telephony companies in southern China), or realpolitik (Singapore Telecom or Telstra in Vietnam), the non-tariff barriers to trade in telecoms services have remained comparatively higher for American and European MNCs.

Intra-Asian Traffic

Another measure of regionalisation is intra-regional telephone traffic. As with trade or FDI, a high percentage of within-region traffic suggests strong commercial links. By 1991, the proportion of outgoing telecoms traffic, generated by Asian countries and destined for other Asian countries was around 56 per cent (Figure 5.2).

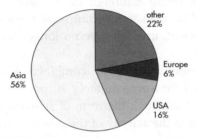

Source: IIC, *Telegeography 1992*.

Figure 5.2 Destination of Asia-Pacific Traffic, 1991

While one should remain wary of overstatement, the relationship between trade and telecommunications traffic across the region appears increasingly strong. Tables 5.4 to 5.7 provide a comparison of the increasing presence and significance of the East Asian countries in each others flows of trade and public switched telephone message traffic. South Korea presents a good illustration of this relationship: taking its top ten trading partners as indicative, between 1985 and 1992, the proportion of South Korea's trade directed towards Asian countries increased from 38.5 per cent (Table 5.4: the bracketed figure in the upper right hand corner) to 52 per cent (Table 5.6). During that time, South Korea's trade outlook had restructured, so that China, Singapore, Indonesia and Taiwan all became significant trading partners, while the previous orientation towards Europe diminished. Over the same time period, again using the ten most popular destinations as indicative, the proportion of South Korea's outgoing telecommunications traffic

Table 5.4
Ten Top Trading Partners for Asia-Pacific Countries, 1985

	Hong Kong (61.7)			China (63.1)			Taiwan		
	Country	US$m	%	Country	US$m	%	Country	US$m	%
1	China	18.0	25.4	Japan	17.5	23.4			
2	US	14.1	19.9	Hong Kong	15.3	20.5			
3	Japan	8.9	12.5	US	7.4	9.8			
4	Taiwan	4.0	5.7	Germany	4.6	6.2			
5	UK	2.8	3.9	USSR	2.7	3.6			
6	Germany	2.8	3.9	Singapore	1.8	2.4			
7	Singapore	2.4	3.4	Australia	1.6	2.1			
8	S.Korea	2.2	3.2	Italy	1.5	2.0			
9	Australia	1.2	1.7	Canada	1.3	1.8			
10	Switzerland	1.1	1.6	France	1.1	1.4			
	Total	57.5	81.2	Total	54.8	73.2	Total		

	Malaysia (61.2)			Singapore (61.5)			Indonesia (62.6)		
	Country	US$m	%	Country	US$m	%	Country	US$m	%
1	Japan	5.5	22.3	US	9.1	18.9	Japan	9.8	36.5
2	US	4.3	17.6	Japan	7.0	14.6	US	4.4	16.4
3	Singapore	4.0	16.3	Malaysia	6.7	14.0	Singapore	2.2	8.2
4	Germany	1.0	4.0	Hong Kong	2.1	4.3	Germany	1.1	3.9
5	UK	1.0	4.0	China	2.0	4.2	S.Arabia	0.7	2.8
6	S.Korea	1.0	3.9	Thailand	1.6	3.2	Taiwan	0.7	2.7
7	Thailand	0.8	3.2	Germany	1.5	3.2	Netherland	0.6	2.4
8	Australia	0.7	3.0	Taiwan	1.5	3.2	Australia	0.6	2.2
9	Taiwan	0.7	2.8	UK	1.4	3.0	UK	0.5	2.0
10	Netherland	0.6	2.3	Australia	1.1	2.3	S.Korea	0.5	1.9
	Total	19.6	79.4	Total	34.0	70.9	Total	21.1	79.0

	South Korea (38.5)			Thailand (59.5)			Philippines (45.2)		
	Country	US$m	%	Country	US$m	%	Country	US$m	%
1	US	20.5	30.9	Japan	3.9	20.4	US	3.0	29.5
2	Japan	16.3	24.6	US	2.9	16.2	Japan	1.7	17.1
3	Germany	2.5	3.7	Singapore	1.4	7.7	Hong Kong	0.5	4.7
4	Canada	2.2	3.3	Germany	0.9	5.2	Germany	0.5	4.5
5	Hong Kong	2.1	3.2	Malaysia	0.8	4.2	Taiwan	0.4	4.1
6	Australia	1.6	2.4	Netherland	0.7	4.1	UK	0.3	3.3
7	UK	1.6	2.4	UK	0.6	3.2	Malaysia	0.3	3.1
8	S.Arabia	1.5	2.3	China	0.5	3.0	S.Korea	0.2	2.8
9	France	1.3	1.9	Hong Kong	0.5	2.7	Singapore	0.2	2.8
10	Malaysia	1.1	1.7	Taiwan	0.4	2.6	Netherland	0.2	2.8
	Total	50.7	74.0	Total	12.6	69.3	Total	7.3	74.7

Source: IMF, *World Trade* (1993).

Table 5.5
Ten Most Frequently Called Countries from Asia-Pacific Countries, 1985

	Hong Kong (73.7)			China			Taiwan (69.2)		
	Country	Calls	%	Country	Calls	%	Country	Calls	%
1	China	5.3	17.8	NA			Japan	4.3	32.2
2	Taiwan	3.7	12.3	NA			US	3.2	24.0
3	Japan	2.9	9.6	NA			Hong Kong	3.1	23.1
4	US	2.7	8.9	NA			Singapore	0.6	4.4
5	Macau	2.5	8.3	NA			Philippines	0.24	1.8
6	UK	1.7	5.8	NA			S. Korea	0.24	1.8
7	Singapore	1.6	5.2	NA			Malaysia	0.21	1.6
8	Philippines	1.1	3.8	NA			Indonesia	0.18	1.3
9	Australia	0.9	3.1	NA			Thailand	0.15	1.1
10	Canada	0.8	2.7	NA			Australia	0.14	1.0
	Total	23.2	81.3	Total			Total	13.04	92.3

	Malaysia (68.7)			Singapore (74.0)			Indonesia (78.6)		
	Country	Calls	%	Country	Calls	%	Country	Calls	%
1	Japan	0.52	16.8	Indonesia	2.5	16.0	Singapore	1.6	39.4
2	Hong Kong	0.36	11.6	Japan	2.3	14.6	Japan	0.62	15.8
3	UK	0.33	10.5	Hong Kong	1.9	12.4	Hong Kong	0.33	8.3
4	Thailand	0.30	9.6	US	1.4	9.1	US	0.31	8.0
5	Australia	0.26	8.4	UK	1.1	6.8	Taiwan	0.15	3.8
6	US	0.22	7.2	Taiwan	0.86	5.5	UK	0.13	3.3
7	Taiwan	0.22	7.1	Australia	0.79	5.1	Australia	0.12	3.1
8	Indonesia	0.15	4.9	India	0.71	4.5	Germany	0.11	2.8
9	India	0.12	3.8	Thailand	0.69	4.4	Malaysia	0.09	2.4
10	S. Korea	0.11	3.6	Philippines	0.40	2.6	Netherland	0.09	2.2
	Total	2.59	83.5	Total	12.65	81.0	Total	3.55	89.1

	South Korea (47.3)			Thailand (57.6)			Philippines (31.9)		
	Country	Calls	%	Country	Calls	%	Country	Calls	%
1	Japan	10.6	37.0	US	2.7	17.9	US	2.3	45.5
2	US	9.4	32.6	Japan	2.3	15.3	Japan	2.3	10.7
3	S. Arabia	2.2	7.6	Singapore	2.1	14.0	Hong Kong	0.50	9.8
4	UK	1.2	4.3	Hong Kong	1.6	10.6	S. Arabia	0.24	4.8
5	Hong Kong	1.2	4.2	UK	0.89	5.8	Singapore	0.19	3.8
6	Germany	0.71	2.5	Italy	0.69	4.5	Australia	0.16	3.7
7	Taiwan	0.45	1.6	Malaysia	0.61	4.0	Taiwan	0.16	3.2
8	Singapore	0.44	1.5	Australia	0.58	3.8	Canada	0.14	2.8
9	Canada	0.33	1.2	Taiwan	0.56	3.7	UK	0.09	1.7
10	Australia	0.30	1.0	France	0.41	2.6			
	Total	26.83	93.5	Total	12.44	82.2	Total	4.35	86.0

Source: AT&T, *The World's Telephones, 1985–86.*

Table 5.6
Ten Top Trading Partners for Asia-Pacific Countries, 1992

	Hong Kong (70.2)			China (70.5)			Taiwan (50.9)		
	Country	US$m	%	Country	US$m	%	Country	US$m	%
1	China	81.2	33.4	Hong Kong	58.0	37.0	US	39.3	25.0
2	US	36.7	15.1	Japan	25.4	16.2	Japan	30.7	20.0
3	Japan	27.7	11.4	US	17.5	11.1	Hong Kong	17.2	11.2
4	Taiwan	15.4	6.4	USSR	6.6	4.2	Germany	7.5	4.9
5	Germany	9.2	3.8	Germany	6.5	4.1	Singapore	4.2	2.7
6	Singapore	8.2	3.4	S. Korea	5.1	3.2	UK	3.5	2.3
7	S. Korea	7.6	3.2	Singapore	3.2	2.1	Australia	3.5	2.3
8	UK	6.8	2.9	Italy	2.8	1.8	S. Korea	3.4	2.3
9	Italy	3.6	1.5	Canada	2.6	1.6	Malaysia	3.4	2.2
10	France	3.3	1.3	Australia	2.3	1.5	Netherland	3.1	2.0
	Total	199.7	82.4	Total	130.0	82.8	Total	115.8	74.9

	Malaysia (68.6)			Singapore (62.6)			Indonesia (68.4)		
	Country	US$m	%	Country	US$m	%	Country	US$m	%
1	Japan	15.8	19.6	US	21.0	15.5	Japan	17.7	31.3
2	Singapore	15.7	19.4	Japan	17.1	13.0	US	7.7	13.6
3	US	13.9	17.3	Malaysia	16.0	11.8	Singapore	4.7	8.4
4	Taiwan	3.5	4.4	Hong Kong	8.0	5.9	S. Korea	3.5	6.2
5	Germany	3.3	4.1	Thailand	5.8	4.3	Germany	3.2	5.7
6	UK	3.0	3.7	S. Korea	5.2	3.8	Taiwan	2.7	4.7
7	S. Korea	2.6	3.2	Germany	4.4	3.3	Australia	2.2	3.9
8	Thailand	2.5	3.1	Taiwan	4.3	3.2	China	2.1	3.8
9	Hong Kong	2.4	3.0	S. Arabia	4.2	3.1	France	1.9	3.3
10	China	1.7	2.2	UK	4.1	3.0	Hong Kong	1.7	3.0
	Total	64.4	80.0	Total	90.1	66.9	Total	47.4	83.9

	South Korea (52.0)			Thailand (66.3)			Philippines (49.0)		
	Country	US$m	%	Country	US$m	%	Country	US$m	%
1	US	37.1	23.4	Japan	17.6	24.0	US	7.1	27.0
2	Japan	31.1	19.6	US	12.1	16.5	Japan	5.8	22.1
3	Hong Kong	6.7	4.2	Singapore	5.8	7.9	Hong Kong	1.6	6.2
4	Germany	6.6	4.2	Germany	3.4	4.9	Taiwan	1.4	5.2
5	China	6.4	4.0	Taiwan	2.9	3.9	Germany	1.4	5.2
6	Singapore	5.0	3.2	Malaysia	2.4	3.3	S. Arabia	0.9	3.4
7	S. Arabia	4.7	3.0	S. Korea	2.3	3.2	UK	0.8	2.9
8	Indonesia	4.2	2.7	UK	2.1	2.9	Singapore	0.8	2.9
9	Australia	4.2	2.6	Hong Kong	2.0	2.7	Malaysia	0.7	2.7
10	Taiwan	3.6	2.2	China	1.6	2.2	France	0.5	2.0
	Total	109.6	69.1	Total	52.2	71.5	Total	21.0	79.6

Source: IMF, *World Trade* (1993); Taiwan statistics from IEU, Taiwan, *Country Report*, 1993.

Table 5.7
Ten Most Frequently Called Countries from Asia-Pacific Countries, 1992

	Hong Kong (77.6)			China (92.3)			Taiwan (68.4)		
	Country	MiTT	%	Country	MiTT	%	Country	MiTT	%
1	China	535.0	47.0	Hong Kong	412.0	64.9	US	82.2	22.3
2	US	80.0	7.0	Japan	56.9	9.0	Hong Kong	64.3	17.5
3	Taiwan	68.0	6.0	Taiwan	47.1	7.4	Japan	55.1	14.9
4	Canada	57.0	5.0	US	30.3	4.8	China	52.9	14.3
5	Japan	45.0	4.0	Macau	24.8	3.9	Singapore	13.4	3.6
6	UK	45.0	4.0	S. Korea	9.3	1.5	Thailand	9.0	2.4
7	Macau	34.0	3.0	Singapore	7.3	1.2	Canada	8.8	2.4
8	Singapore	34.0	3.0	Australia	6.2	1.0	Philippines	8.5	2.3
9	Australia	34.0	3.0	Germany	5.2	0.8	Malaysia	8.3	2.2
10	Philippines	34.0	3.0	Canada	4.9	0.8	Australia	6.8	1.9
	Total	966.0	85.0	Total	604.0	95.3	Total	309.3	83.8

	Malaysia (80.8)			Singapore (78.0)			Indonesia (70.0)		
	Country	MiTT	%	Country	MiTT	%	Country	MiTT	%
1	Singapore	97.3	44.9	Malaysia	110.0	26.7	Singapore	28.8	24.4
2	Japan	16.3	7.5	Indonesia	40.0	9.7	US	12.5	10.6
3	US	12.7	5.9	US	35.0	8.5	Japan	12.3	10.4
4	Australia	12.0	5.5	Hong Kong	34.0	8.3	Australia	8.5	7.2
5	UK	11.2	5.2	Japan	34.0	8.3	Hong Kong	8.3	7.0
6	Hong Kong	10.4	4.8	UK	22.0	5.3	S. Korea	6.0	5.1
7	Taiwan	9.2	4.2	Thailand	19.0	4.6	Taiwan	5.9	5.0
8	Thailand	7.4	3.4	Australia	17.0	4.1	Malaysia	5.2	4.4
9	Indonesia	6.0	2.8	India	14.0	3.4	UK	4.1	3.4
10	India	4.5	2.1	Philippines	12.0	2.9	Germany	3.4	2.9
	Total	187.0	86.3	Total	337.0	81.8	Total	95.0	80.4

	South Korea (53.6)			Thailand (60.3)			Philippines (38.1)		
	Country	MiTT	%	Country	MiTT	%	Country	MiTT	%
1	US	93.5	30.6	Japan	23.1	17.4	US	53.5	39.4
2	Japan	80.1	26.2	US	20.5	15.5	Japan	21.1	15.5
3	Hong Kong	14.6	4.8	Singapore	13.4	10.1	Hong Kong	11.8	8.7
4	Philippines	12.6	4.1	Hong Kong	11.7	8.8	Canada	7.3	5.4
5	China	10.8	3.5	Taiwan	9.2	6.9	Australia	6.0	4.4
6	Germany	7.5	2.5	UK	6.6	5.0	Singapore	4.4	3.2
7	Taiwan	7.3	2.4	Australia	5.2	3.9	Taiwan	3.9	2.9
8	UK	6.6	2.2	Germay	4.6	3.5	S. Korea	3.3	2.4
9	Australia	5.7	1.9	China	3.9	2.9	S. Arabia	3.0	2.2
10	Singapore	5.7	1.9	France	3.5	2.6	UK	2.5	1.8
	Total	244.4	80.1	Total	101.7	76.6	Total	116.8	85.9

Source: IIC, *Telegeography* (1993).

directed towards Asian countries increased from 47.3 per cent (Table 5.5) to 53.6 per cent (Table 5.7). Similarly, by 1992, six of the top ten partners were Asian (Japan, Hong Kong, the Philippines, China, Taiwan and Singapore).

However, telecommunications traffic flows are often also an indicator of strong social links, and thus may indicate the potential for trade before trade links have been established. From this perspective, in 1985 one would have expected South Korea's Asian trade links to grow as the result of where its interests lay. (The emergence of China as a trade and telecommunications partner for South Korea has obviously been constrained by political realities.)

Similarly, the rate of increase in telecommunications trade between China and Taiwan recently is an indication of the restrictions placed upon Taiwanese contact with the mainland.[5] Prior to 1985, traffic between the two registered only several thousand calls a year. Following Taiwan's authorization of indirect two-way communication in June 1989, the number of calls began to increase rapidly, with 1.8 million calls registered from China to Taiwan that year, and 24.9 million calls from the mainland in 1993. In May 1994, the mainland became the number one calling destination for the Taiwanese, surpassing the United States and Japan. For Chinese callers, Taiwan has become the second most called locale after Hong Kong and ahead of Macau.[6]

Telephone traffic flows may therefore provide a broader picture than macroeconomic indicators such as imports and exports, in this new world of services trade. For example, looking at Hong Kong's trade and telecommunications relationships in 1992: telecommunications with Canada, Macau, Australia and the Philippines (Table 5.7) displace Germany, South Korea, Italy and France from their trade rankings (Table 5.6) respectively due to emigration, gambling, emigration and domestic labour. Hong Kong, as a result of its colonial heritage, has acted as a primary Asian trade hub for European companies. Telecommunications message traffic in the early 1990s, however, suggests that the territory's interests increasingly lie elsewhere.

Two caveats must be borne in mind when reading the above tables. The first is that in comparing the 'most frequently called countries', the first table, Table 5.5 (1985) uses the number of calls — that is, the number of discrete messages — as the basis for assessment, while Table 5.7 (1992) uses 'MiTTs', or the *number of minutes* of telecommunications traffic. Admittedly, this is a little like comparing apples and oranges, and for this reason, all comparisons have been done on the basis of a percentage relationship.

Secondly, statistics for *outgoing* calls are calculated from the place of billing rather than the place of origination (of the call). This will present statistical problems where calls are regularly reversed to take advantage of substantially cheaper tariffs in one direction. (This has become a particular problem for the US, which is now running a significant deficit with most countries.[7] Because, for example, it is much cheaper to call from the US to the Philippines than the other way, many calls from the Philippines are reversed. This means that although the call is made from the Philippines, it is registered as a phone call made from the US. It is estimated that 10 per cent of the total outgoing minutes of traffic from the US in 1991 were actually *incoming* minutes. In some countries the percentage of reversed or 'calling card' traffic is significantly higher, up to 90 per cent. The extent to which this may distort the above statistics is unclear.

Multinational Corporations: The New Mercantilists?

Estimates made by the ITU in 1992 were that, of the US$415 billion generated in worldwide telecommunications service revenues, the world's top 2500 multinational corporations contributed US$50 billion (ITU, 1994). Of this, US$10 billion was generated in Asia.

The international services market in telecommunications has been largely created by this top tier of multinationals and their demands for advanced networking services in the countries in which they do business. To integrate and coordinate their businesses across different countries and different continents, this high-value group of customers requires operators to supply services from a single source and of a consistent standard around the world.

As a result, in 1992, when the globe's inhabitants generated more than 40 billion minutes of international telephone traffic, almost 80 per cent of this was carried by just 15 companies. Growing at approximately 17 per cent per year over the last decade, international public switched traffic is not only the fastest growing part of the telecoms business; it is also the most profitable. It is for this reason that the multinational telephone companies have been attempting to slowly but surely prise international traffic away from foreign telephone companies and into their own networks.

Regulatory controls have so far kept the major carriers out of many overseas markets, but, as with service companies elsewhere, they have been able to follow their customers abroad, providing global account management functions, interfacing with foreign carriers and offering systems integration expertise. In developments which have been interpreted as reflecting the global movement towards regional trade blocs — but which appear eerily reminiscent of the mercantile shipping empires of the early industrial revolution — these dominant telecommunications carriers have begun to group together so as to provide regional, or global, telecommunications services, and to integrate foreign countries into their global information networks (Table 5.8).

Modern telecommunications networks have begun to form across the Asian region as the result of two compelling, and at times competing, forces:

Table 5.8
Global and Regional Telecommunications Groupings

Carrier(s)	Partnerships	Network plans	Service plans
AT&T	AT&T Istel (UK), AT&T JENS (Japan), and NCR	European backbone through AT&T Istel	Clear channels, packet switching, frame relay, messaging.
BT/ MCI	Syncordia; BT North America; MCI Communications	Global backbone.	Clear channels, virtual private network, outsourcing.
C&W	Owns, wholly or partially, public network operators in more than 40 countries.	'Global Digital Highway (includes European backbone).	Global managed data service.
Eunetcom/ Sprint	Deutsche Bundespost Telekom, France Telecom and Sprint (USA).	Originally planned as European backbone, Sprint's association increases their reach.	Packet switching, frame relay, managed bandwidth, virtual private network.
FNA (a)	Headed by MCI; includes Belgacom, France Telecom, Hongkong Telecom, Italcable, KDD, Mercury Communications, Singapore Telecom, Stentor, Telstra and Telefonica.	Global backbone.	Leased lines initially; other services under discussion.
GEN (b)	BT, Deutsche Bundespost Telekom, France Telecom, STET (Italy), and Telefonica	Backbone between member countries.	Not marketed directly to end-users.
GNP (c)	Collaboration involving AT&T, BT, Deutsche Bundespost Telekom, France Telecom, KDD and Telstra.	Backbone between members.	Not marketed directly to end-users.
Metran (d)	Collaboration involving 25 PTTs	SDH backbone.	Not marketed directly to end-users.
Temanet A/S	Telecom Denmark and Maersk Data A/S	35-node network.	Leased lines.
Unisource NV (also a part of WorldPartners)	PTT Telecom (Netherlands), Swedish Telecom and PTT Telecom (Switzerland).	European backbone (negotiating inteconnection with nertworks in Japan and US).	Leased lines, data services, messaging, net. management.
WorldPartners	AT&T, KDD and Singapore Telecom. (Telstra, Korea Telecom, Hongkong Telecom, Telecom New Zealand, Unitel Canada and Unisource are 'associate members')	Global backbone	Leased lines, virtual private voice/data network, frame relay.

Note: (a) Financial Network Association
(b) Global European Network
(c) Global Network Project
(d) Managed European Transport Network

1. global multinational corporations (or collaborations) adding an Asian operation to their network;[8]
2. regional multinationals ('national champions') expanding to capitalize on geographic, or related, competitive advantages.

The idea behind the 'national champion' strategy has been that, through joint ventures with MNCs, the state gains access to development and technology transfer. The state is also able to retain a degree of indigenous control while providing the environment for local companies to become substantial players. The first stage in this strategy is to monopolise the sector and build large institutional interests (i.e., a fairly standard infant industry approach). The next stage is to build on the monopoly player and, then finally, for the monopolist to become an international (initially, regional) player.

As a result, local telecommunications equipment and network management companies in the East and Southeast Asian region have recently begun to expand and diversify into regional markets, including Australia, the ASEAN countries, Indochina and China, but at this stage in their development they stand little chance of competing internationally or offering global services. Singapore Telecom's subsidiary, STI, has been signing agreements in the Philippines, Indonesia and Thailand as well as Indochina and China; Technology Resources (TR) of Malaysia has entered the Indochina market and is actively seeking entry elsewhere; the Thai group, Shinawatra, has similar ventures, while the CP Popkhand group has begun to move into China,[9] along with a number of Hong Kong ventures. Australian carrier Telstra has stated its intentions to be the dominant Asian carrier.

In these areas, ASEAN's or Northeast Asia's companies can develop synergies, or business complementarities, which have as much to do with business, ethno-linguistic or governmental networks, as they do with more standard macroeconomic indicators such as flows of foreign direct investment, or industry sourcing programmes. As local firms begin to expand, both home and host governments become more attuned to their objectives, reinforcing social interests (such as migration flows), and providing a basis for regional cooperation, economic liberalization, and the regional coordination of sectoral policy (such as the deregulation of the telecoms markets).

The Integrated Services Digital Network (ISDN)

As firms seek to expand their enterprises across national boundaries, they will increasingly look to employ computer-communications networks to their own needs. It is this that is promoting the trend away from nationalization of telecommunications infrastructures towards private companies and the introduction of competition. In those nations where planning for transmission infrastructures is only recently under way, for example Indonesia and Thailand, the pressures from user groups (more often than not multinational users) will be responsible for forcing a shift to privately owned telephone systems. Where infrastructures already exist, as in Malaysia and the Philippines, the choice of future technology is being debated, for example will integrated services digital network (ISDN) provided by a national carrier dominate or will a confederation of user-operated telecommunication networks that can interconnect with the public system be the preferred option.

ISDN constitutes a series of technical protocols permitting the transmission of all possible forms of messages (voice, data, video, graphic) on one single telephone line, in separate or integrated forms. In other words, the same line may transmit cable TV broadcasts and phone calls simultaneously, and/or serve as a picturephone or videoconferencing device integratively with data transmission.

Many commentaries speculate that ISDN will further reduce national governments' control over information flows between MNC operations because ISDN will allow computers, telephones, data bases and other information services throughout the world to communicate with each other directly, without translation from analogue to digital. ISDN is based on the premise of standardizing all voice, data, telex, and even video telecommunications links through all-digital transmission modes. As such, ISDN would provide for interconnection of computers worldwide in the same way the present telephone direct-dial system links countries and continents with standardized telephone dialling codes, billing and technical transmission standards.

The main problem with this outlook is that ISDN is a series of telecommunications protocols, *not* computer protocols. If the stricter regulatory regime for telecoms were to predominate in the merging arena of telecommunications and computing technology, less international trade in services would result, all other factors being equal. Indeed, the way in which the regulatory apparatus in each country accommodates the merger of computing and telecoms will have obvious consequences for international trade in services, for the politics of the telephone are the politics of trade. Standards setting for information services such as ISDN that are too strict can impede international trade in services. Standards setting can be used by governments to prevent foreign competition or to favour preferred domestic suppliers. If the PTTs of the world provide their own forms of ISDN, there will continue to be little opportunity for trade in telephones and terminal equipment.[10] (See also Chapter 7.)

Network Facilities

There remains a perception that domestic and international telecommunications services are quite distinct entities, fuelled by both the national PTTs, which would like to maintain their 'natural' monopolies, and the international carriers, which would like to maintain an international cartel. By this view, domestic telecommunications is essentially a question of development or state-induced processes of reform, while international telecommunications services can be paralleled to the airlines industry: a basic amount of infrastructure is required (airport landing facilities) and airlines, — be they state-owned or privately owned — are granted rights, for a fee, to land their traffic at the international gateways. State ownership of an international telecoms carrier is then much like ownership of an international airline carrier, having as much to do with ego and posturing as with strategic importance. There are problems with this interpretation of the provision of telecommunications services.

To understand the provision of international telecommunications services and appreciate the multinational alliances being formed, one needs to recognize the trading regime within which the carriers are working. As the liberalization of domestic telecommunications industries continues, the business of providing international telecommunication services is going through a paradigm shift. (See Staple, 1992.) Traditionally, national telephone companies argued that telecommunications services did not constitute trade: international calls were jointly provided services, the product of a joint investment by two or more countries in a common infrastructure (transoceanic cables and, since the 1960s, satellite distribution) connecting the countries. Theoretically, voice and data messages were handed off at the midway point between the sending and receiving country. This became known as the 'half-circuit regime', with international service provided by national carriers who interconnected their half-circuits so as to provide end-to-end service.

Even where competition existed, the existing trade regime made it physically impossible for a carrier from one country to pick up and land traffic in a second country without a carrier from the latter country supplying facilities. Facilities-based, the carriers, in effect, jointly owned the international cable and satellite systems — the global telecommunications infrastructure.[11]

Technology and liberalization, however, are leading to the existing order being challenged from several different directions. These can be summarized into three:
1. opportunities arising from the development of a regional infrastructure (best illustrated in the provision of submarine cable facilities),
2. the emergence of private suppliers (across Asia this has occurred predominantly in the provision of satellite distribution facilities),

3. the emergence of traffic (and business) hubs within the region, based upon the provision of value-added, and private, network services.

The rest of this chapter examines each of these three aspects and their impact upon the trade in telecoms services regionally.

The Cables: An Information Infrastructure

Since 1989, at least 20 international fibre optic cable systems have either started service or will, by 1998, come into service, in the Asia-Pacific region (Figure 5.3; Table 5.9). The carriers were committed to spending in excess of $US6 billion on international fibre in the region.

The contracts to lay transoceanic cable systems have historically gone to a relatively small group of companies, so that the submarine cable club, much like the satellite club, was run as a closed shop. However, liberalization of domestic industries and the ability to trade in these services, has allowed other names to appear. Among the relative newcomers to the Asian region's cable systems are France Telecom International in the Sea-Me-We 2 grouping; America's Nynex, founder of the FLAG system which roughly parallels Sea-Me-We 2; and US Sprint Communications, one of the nine Aspac partners.

Fibre optic cables offer greater capacity and faster response over short-to-medium distances for all traffic, and over long routes (such as across the Pacific) for high-speed data. This has lead to a resurgence in the usefulness of submarine cables and a shift away from a reliance on satellite links for data transmission. As unit costs drop, the international carriers are racing to become the first to develop optical fibre links, primarily because of the long-term strategic advantage that it gives to the cable owner.[12]

The prize sought by these groups is the operation of hubbing centres throughout the world, picking up fees for transit traffic where it passes through shore equipment and having enough extra capacity to lease or sell it to other operators. This represents a dramatic shift of mind-set: sub-oceanic cable systems are no longer viewed as public or semi-public utilities. Rather, they are lucrative tollroads whose very creation generates more traffic and thus more revenue.

The Satellites: Footprints on Asia

In early 1993, it was reported, somewhat ironically, that Taiwan and China were planning to launch a Hong Kong partnership to design and build a regional telecommunications satellite. With Japanese and Singaporean in-

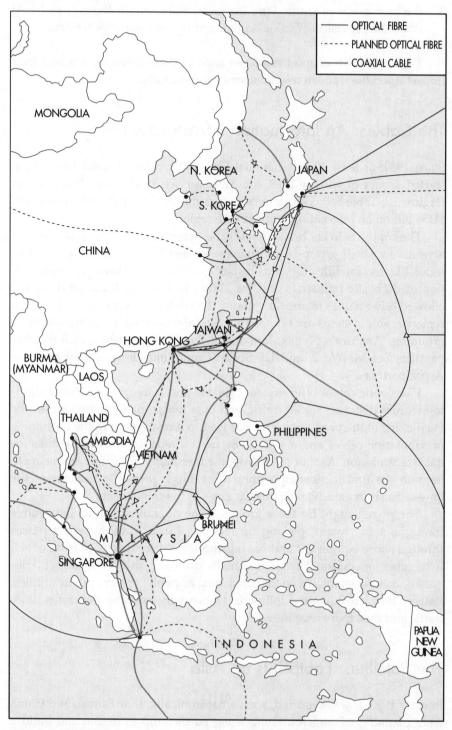

Figure 5.3 Cable interlinkages in Asia-Pacific (source: Hongkong Telecom, January 1995)

Table 5.9
Fibre Optic Submarine Cable Systems in the Asia-Pacific Region

Cable System	Section	Length (km)	channels	# of carriers	Date in service
TPC-3/	Japan — Guam (1)	9070	(1) 3191	30 carriers	April 1989
HAW-4	Hawaii (2) — US Mainland	4230	(2) 2810	15 countries	
TPC-4	Japan — US Mainland — Canada	9800	15120	50 carriers 34 countries	Nov 1992
NPC	Japan — US	8380	17010	32 carriers 19 countries	May 1991
TPC-5	Japan — US — Guam — Hawaii	25000	120960	48 carriers 34 countries	1995–96
HAW-5	US Mainland — Hawaii	4400	15120	29 carriers 16 countrie	Jan 1993
PacRimEast	New Zealand — Hawaii	7700	7560	24 carriers 15 countries	Mar 1993
PacRimWest	Guam — Australia	7500	7560	29 carriers 17 countries	Dec 1994
TASMAN-2	Australia — New Zealand	2200	15120	38 carriers 25 countries	Mar 1992
G-P-T	Guam — the Philippines — Taiwan	3750	3780	28 carriers 14 countries	Jan 1990
HONTAI-2	Hong Kong — Taiwan	730	5670	9 carriers 6 countries	Aug 1990
H-J-K	Japan — Hong Kong (1) S. Korea (2)	4590	(1) 3780 (2) 3780	29 carriers 13 countries	July 1990
T-V-H	Thailand — Vietnam — Hong Kong	–	–	–	end 1995
China-Japan	Japan — Shanghai	1250	7560	5 carriers 3 countries	Dec 1993
China-Korea	Qingdao — Taean	570	15120	–	Dec, 1995
R-J-K	Russia — Japan Japan — S. Korea Russia — S. Korea	1700	7560	–	1995
APC	Japan — Taiwan — Hong Kong — Malaysia — Singapore	7520	7560 per pair	38 carriers 23 countries	Sept 1993
APC-N	Japan, Korea, Taiwan, Hong Kong, Philippines, Malaysia, Thailand, Indonesia, Singapore	–	–	–	1996–97
FLAG	Korea and Japan via Singapore, Hong Kong, Indonesia to the UK	–	–	–	Dec, 1996
ASEAN Cable (AOFSCN)	Philippines — Malaysia — Brunei (1) — Singapore (2) Malaysia (3)	4500	(1) 15120 (2) 7560 (3) 11340	(1) 27 carriers 16 countries (2) 23 carriers 12 countries (3) 22 carriers 12 countries	(1) Feb 1992 (2) Dec 1991 (3) Feb 1992
SEA-ME-WE2	Singapore and Indonesia to France via Egypt	14000	7680 per pair of fibre	53 carriers 41 countries	mid 1994
SEA-ME-WE3	Singapore and Indonesia to France via Egypt	in planning	in planning	17 carriers 11 countries	in planning

terests also expected to participate, the total capital invested would be around US$100 million. This company would be the third regional satellite company to be based in Hong Kong in the space of less than five years.

Yet, prior to the 1990s, the Asia Pacific had only one regional satellite system — the Indonesian-owned Palapa — which did not enjoy full capacity because the Indonesians could not persuade their neighbours to purchase satellite services. A few years down the track and most of the larger countries in the region had either launched or made elaborate preparations to send up their own satellites, including Indonesia, Japan, South Korea, Singapore, Hong Kong, China, Thailand and Malaysia. By late 1995, the number of satellite transponders available for Asia will have almost tripled from that available in 1992 (Figures 5.4 and 5.5).[13]

This has led to speculation of an impending price-war due to the resulting over capacity. Indeed, transponder prices have been falling, but this has not been due to price competition so much as utilization and accessibility.[14] The issue over the Asia-Pacific (at this stage) is not over-capacity but rather usage, strategic positioning and — as with so much else to do with business in the region's uninstitutionalized environment — government connections, for the appropriate regulatory requirements and business demand.[15]

With common carrier telecommunications still the most common application for satellite equipment, fibre optic submarine cables are challenging this traditional satellite market by providing virtually unlimited bandwidth. However, satellite communications can now be expected to diversify. The satellite market will remain strong and grow in the areas of ISDN and, by the late 1990s, telephony will likely be overtaken by point-to-multipoint broadcast communications, for example, television distribution. Across Asia, the demand for satellite services will be accentuated by the ability to provide services to rural communities[16] and through roaming services which

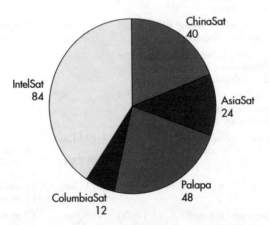

Figure 5.4 Regional satellite transponders (total 208) for Asia, 1992.

Information Highways and the Trade in Telecommunications Services 133

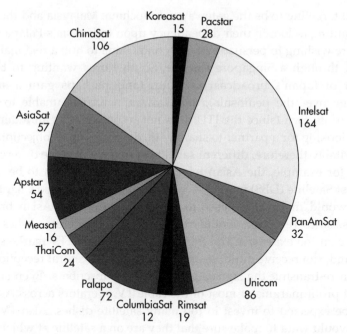

Figure 5.5 Regional satellite transponders (total 685) for Asia, 1995.

Table 5.10
Satellite Communications

Frequency allocation	Uplink/ Downlink	Usage
C-band	6/ 4 GHz	long-haul TV and telephony communications (frequency is similar to microwave terrestrial for analog)
Ku-band	14/ 11 GHz	domestic TV; VSAT; portable satellite and TV trucks
Ka-band	30/ 20 GHz	data transmission
L-band	1500–1700 MHz	signalling and control; navigation

Source: Lido Organization.

will continue the bridging mechanism role that mobile telephony has played in the region while telecommunications infrastructures are slowly built.[17]

This does not make all of the domestic and regional satellite systems now being considered economically viable. The proliferation of satellites has to do with issues of national status, strategic bargaining over satellite slots, overspill and transborder information flows, and positioning for, and posturing over, regional hubbing and sub-regional hubbing. Thus, we find

Thailand targeting to be the gateway to Indochina; Malaysia and the Philippines wanting to loosen their dependency upon Indonesia's Palapa system; Singapore wishing to persuade ASEAN and China to hub a regional mobile network through a Singapore satellite; South Korea wanting to be independent of Japan's broadcast satellites (and perhaps gain a strategic advantage over the peninsula); and Taiwan, which is unable to have a satellite of its own (since the ITU does not recognize it as an independent nation), looking for a partner to share with, despite China's objections.

Inevitably therefore, different satellite systems will serve different markets. If, for example, the Asian television market was seen to be a Direct Broadcast Satellite (DBS) market — which is unlikely (see also Chapter 6) — then it would make little sense to have competing pan-Asian broadcast satellites, since reception would require multiple satellite dishes, which would be far too expensive and for too unwieldy. Cable TV markets, on the other hand, can receive multiple satellite signals at a central reception point and then re-transmit the signals to individual subscribers. Even so, given the tight profit margins of most of the cable TV operators across Asia, they cannot be expected to invest in too many satellite dishes. Also TV broadcasters would want to make sure that they are on a satellite at which people will aim their satellite receivers. Hence, the issue of which satellite becomes the 'hot bird' — the satellite likely to have dishes pointed towards it and the satellite which the broadcasters therefore will be most keen to rent transponder space on (Table 5.11).

Equally, if the broadcast satellite is to be much more focused (for example, its target market is Indochina, rather than pan-Asia), then the requisite 'footprint' can be much more focused, satellite receivers can be smaller (and cheaper) and different transponders can be utilized or rented — say, Ku-band instead of C-band (Table 5.10). If the satellite communications are to be used for domestic telephony or data communications (for example, a VSAT network) then Ku- rather than C-band transponders will be deployed; if for high-speed data transmission, then Ka-band.

One interesting aspect of this development has been the emergence of private companies supplying satellite services, on a global, regional or domestic basis (Table 5.11). Not so long ago, international satellite communications were run as a cartel, with services provided by cooperatives such as Intelsat. Asian, American and European companies — some with no previous experience in the industry — are now investing in the facilities through which they can provide services or lease capacity. Again, what this suggests is that within the framework of the global carriers or consortiums, there have emerged opportunities for local companies to exploit their knowledge of the region and their business relationships; companies such as Hutchison and Pacific Century (run respectively by Hong Kong tycoon Li Ka-shing and his second son Richard Li), Shinawatra, and

Table 5.11
Satellites Over the Asian Region

System	Proprietor	Service	Cover Area	Launch Date	Transponders
Koreasat	Korea Telecom	fixed satellite service & DBS	Korea, North China, Japan	1995	15Ku
Chinasat Dongfanghong	China's DGT	fixed satellite service & DBS	China, Northeast Asia	4: 1988/3 5: 1988/12 6: 1991 3-OD: 1993 3-A1: late1994 3-A2: early 1995	4: 4C 5: 12C 6: 24C 3-OD: 6Ku; 12C 3-A1: 24C 3-A2: 24C
AsiaSat 1&2	Asia Satellite Telecoms	fixed satellite service	South & East Asia + Mid-East	1: 1990/4 2: 1994	1: 24C 2: 24C; 9Ku
Apstar 1&2	APT Satellite	fixed satellite service	East Asia	1: 1994/6; 2: 1994/12	1: 18C & 4Ku 2: 24C & 8Ku
Measat	Binariang	fixed satellite service	Malaysia, Phil., Indonesia	1994	12C; 4Ku
ThaiCom 1&2	Shinawatra Satellite	television, data, telephone	Malaysia, Thai, Sing, Indochina	1: 1993/12 2: 1994	10C; 2Ku 10C; 2Ku
Palapa-B1 Palapa-B2P Palapa-B2R Palapa-B4 Palapa-C	PT Pasifik Satelit Nusantara	video, voice, data	Southeast Asia	B1: 1983/6 B2P: 1987/3 B2R: 1990/4 B4: 1992/5 C: 1996	B2P: 24C B2R: 24C B4: 24C
TRDS-3 (ColumbiaSat)	Columbia Comms	video, voice, data	East Asia - Pacific	1991	12C
Inmarsat II&III	Int. Maritime Sat. Corp	voice, data, fax	global	II: 1992 III: late1995	250 circuits; 1000 circuits
Gorizont/ Express	Rimsat	video, voice, data	global	G: 1993/11; E: 1994	G: 6C; 1Ku; E: 10C; 2Ku
Pacificom	TRW	video, voice, data	Asia Pacific	1997/98	8C; 11Ku
Unicom I&II	Unicom Sat. Corp.	video, voice, data	I: Pacific; II: E. Asia-Europe	1995	36C; 50Ku
PanAmSat (PAS-2)	Alpha Lyracom	video, voice, data	Asia Pacific	May 1994	16C; 16Ku
Intelsat VII (701/703) & VIII (801/802)	Intelsat	video, voice, data	global	701: 1993; 703: 1994; VIII: 1996	26C,10Ku; 38C,6Ku
Pacstar 1&2	Pacific Satellite	video, voice, data	Asia Pacific	mid-1994	12–24C; 4Ku

Source: *APT Journal*, January 1993, p.21; Cooperman, William and Dennis Connors, 'Communications Satellites for the Pacific Region: 1994 Status Report', PTC94, pp.654–61.

CP Popkhand. What remains an interesting question at this point in time is whether the dominant satellite telecommunications service providers to the region will be local entrepreneurs like the Li's, global players such as a Rupert Murdoch, or regional alliances grouping together say, Japanese, Chinese and Korean capital, or Malaysian, Indonesian and Thai industry.

VSATs

While private network opportunities in the East Asian region remain, at this stage, comparatively small, the response to very small aperture terminal (VSAT) services is perhaps indicative. VSAT operators in Indonesia, Thailand and the Philippines have enjoyed steady growth in their networks, despite the relatively high lease rates and recent entry of new operators. This has been driven by the business community, for whom satellite services represent an expedient solution to unavailable terrestrial connections.

Additionally, many of the original VSAT data-only service providers are now developing both voice and value-added services in a variety of formats. Nationwide paging networks utilizing VSAT-based backhaul trunks have been established in the Philippines, Indonesia and Thailand, and there is speculation as to the application of this technique for PCN-based systems across rural areas.[18]

For rural telephony across Asia, VSAT systems provide a means of introducing services much faster than will be the case with a traditional wireline network, because they are distance insensitive. In fact, as has been noted elsewhere, development of a 'generic satellite/wireless drop-in telecommunications package' (Zita, 1993) offers one vehicle for sustained market growth through PSTN by-pass. Facilities service providers who are able to offer shared-use earth stations can reduce user tariffs and thereby address a much broader market base. Further use of satellite services, such as for value-added networks, can also be expected to open up market segments for the satellite industry. These will likely include dedicated leased circuits for data, the consolidation of dial-up access and data broadcast applications via satellite bearer circuits employing data compression.

Broadcast

A further issue brought about by the application of new satellite technology has been 'signal spillover' and what this means for national sovereignty, on the one hand, and programme broadcasters, on the other. With unscrambled signals making programmes available for free reception, satellite services affect local broadcasters' interests. This is particularly the case when people use signal spillover as a provider of external information. For example, in Taiwan the 70,000 or so satellite dishes installed are known locally as 'big ears' (C-band) or 'small ears' (Ku-band) according to their size. They were used initially by the Taiwanese to watch Japanese NHK programmes about mainland China, and at one point Taiwanese viewers actually asked NHK to broadcast more programmes and news about China.[19] This issue is necessarily complicated when different ideologies are involved.

These questions became relevant in Asia with the launch of AsiaSat-1. AsiaSat-1's ownership is split almost equally among the UK's Cable & Wireless, Hong Kong's Hutchison and the Communist Chinese overseas investment arm, the China International Trust and Investment Corporation (CITIC). China's ownership, and its scheduled resumption of sovereignty over Hong Kong, raised concern that the satellite might be used as a propaganda outlet.

While AsiaSat-1 has been the most newsworthy of the new satellites with its STAR TV service, it is not the only facility available (Table 5.12). CNNI's programmes are now beamed over the Superbird (Japan), Arabsat 1C (Middle East), Intelsat F8 (Asia Pacific) and Palapa B2P satellite. The Indonesians, who held a 30 per cent share of the regional satellite services market, are losing ground to the new 'hot birds' — AsiaSat-1 & 2 and Apstar-1 & 2 — as the programmers look to buy transponder space on these new satellites at the expense of Palapa. A further loss for the Indonesians has been the migration of the Thais to ThaiCom. The Thais were Palapa's second largest user group, occupying seven-and-a-half transponders.[20]

The APT Satellite Co. which operates the Apstar satellites is jointly owned by mainland Chinese companies — China Telecommunications and Broadcast Satellite Corp, China Yuan Wang Group and Ever-Victory System Co. — and two Taiwanese companies, including Chia Tai International Telecommunications, which are investing US$109 million. This grouping links China Development Corp with Taiwan's ruling Nationalist Party and the powerful Ruentex group.

The impact in the field of media services has been to shift choice in programming and distribution away from the national broadcaster, which up until now has enjoyed a monopoly position on the service, and into the realm of international media groups. From this perspective, the provision, or financing, of regional satellite services (such as provided by Turner International or News Corporation), or the formation of regional alliances (Tables 5.11 and 5.12), has been the result of multinational capital looking to exploit the limitations of the state system. (See also Chapter 6.)

Table 5.12
Satellite Broadcasters Over Asia

	Owners	Launch	Transponders	Channels	Reception area
AsiaSat-1	CITIC (China), C&W (UK), Hutchison Whampoa(HK)	April 1990	24 C-band	MTV*/ BBC*/ STAR Plus/ Prime Sports/Movie Channel/ Chinese Channel (STAR), Zee TV,Yunnan TV-1, Guizhou TV-1, CCTV-4, Myanmar TV, TV Mongol	China, Japan, Taiwan, HK, ASEAN Indochina, S.Asia, W.Asia
Palapa B2P	Indonesian Govt	Mar 1987	24 C-band	CNNI, HBO, ESPN, ABN, RCTI, RTM-1, TV3, ATVI, CFI, SBC, TVBI, ABS-CBN, AN-Teve, GMA, People's Net, TPI, SCTV, Nine Net	ASEAN, S.China, Indochina, HK, Taiwan, E.India
Intelsat 508	International satellite consortium	Mar 1984	24 C-band	CCNI, ESPN, Network Ten, ABC, Channel 9, World Net/ C-SPAN, Deutsche Welle TV, NHK, RFO Tahiti, Nine Net, TBS, Fuji TV	Japan, Taiwan, Phil., S.China, HK, Australia
ThaiCom1	Shinawatra (Thai)	Dec 1993	10 C-band 2 Ku-band	BC News, IBC Entertainment, IBC Sports, IBC Thai Variety, IBC HBO, Thai Ch-3, Thai Ch-5, Thai Ch-7, Thai Ch-9, Thai Ch-11, Cambodian TV	Thai, Burma, HK, Indochina, Sing., W.Malaysia, Taiwan, S.China, Korea, Japan
Apstar-1	(a)	June 1994	18 C-band 4 Ku-band	CNNI, Time Warner, HBO, ESPN, ATVI, Discovery, Viacom, TNT, Cartoon Net, TVB, NBC.	China, Japan, Korea, Mongolia, HK, Taiwan, ASEAN, Indochina
AsiaSat-2	CITIC (China), C&W (UK), Hutchison Whampoa(HK)	Dec 1994	24 C-band 9 Ku-band	V/ Star Plus/ Prime Sports/ Chinese Channel/ Zee TV/ Children's Channel/ Asian Business Channel/ Movie Channel/ Asian Movies (STAR)	China, Japan, Korea, Taiwan, HK, ASEAN, Indochina, S.Asia, W.Asia
Rimsat-1	Tajudin Ramli (Malaysia), James Simon (US)	Nov 1993	7 C-band 1 Ku-band	Taiwan TV, Sun TV	S.China, Japan, Taiwan, Korea, HK, ASEAN, Indochina, S.Asia
Apstar-2	(a)	Dec 1994	26 C-band 8 Ku-band	CNNI, Time Warner, HBO, ESPN, Turner, Viacom, TVBI	China, Japan, Taiwan, Korea, Mongolia, HK, ASEAN, Australia, India, W.Asia

* These channels have been subsequently dropped from the services; (a) Group comprised of: China government consortium, Kwang Hua Development (Taiwan), CP Group (Thailand), Singapore Telecom.

Regional Telecommunications Hubs (RTHs)

Large MNCs are the target market for commercial and telecoms hubs, given that they generate a very large percentage of the total traffic in private international telecommunications networks.[21] Hubs provide both a controlling focus for regionally connected customers and a transit service for traffic from other hubs on the network. They include private and virtual lines, and they will usually be locally connected to a service such as a data processing centre to attract the traffic of 'offshore' users. In Asia, some 60 per cent of private communications hubs are in Hong Kong,[22] although many companies have now begun to use Singapore as well to capture traffic from the fast growing Southeast Asian region. Tokyo, by way of contrast, has not functioned as a centre for RTHs in Asia, primarily because of its high cost location.

RTHs actually account for a relatively small percentage of overall business traffic,[23] but they do generate a significant amount of traffic on the public switched network. And, of course, regional headquarters and regional services attract ancillary industries to develop around them, and enhance exports of other service industries, including banks and financial institutions, accountancy, legal and research services, and others.

With the regional services market growing, international trade in services and increasing foreign direct investment in service industries has not only made Asia an attractive trade or investment destination, it is changing the relative balance within Asia itself. As a result, a number of other Asian centres have attempted to establish themselves as the place for RTHs (or RHQs), by offering financial or non-financial incentives. There has also been an attempt to capitalize on the uncertainty surrounding Hong Kong's continued viability after 1997. Taiwan, for example, has a long-term goal to become a financial and telecommunications centre within Asia. To this end, they have begun to deregulate the value-added and international value-added services market, and to introduce competition, so as to reduce tariffs. Similarly, Thailand is making a substantial effort to improve its domestic and international telecommunications services, with the aim of becoming a regional telecommunications and data processing hub.

Value-Added and Information Services

The provision of diverse, high-quality, value-added services represents increased access and integration into the global economy, which in turn requires world-class, and world-competitive, efficiency. These are the benefits of expanding international intracorporate networks. The countries with the most cost-effective and advanced telecommunication services are most likely to attract the corporate and regional headquarters of global companies which, as we have seen, dominate telecoms traffic. Thus, one argument regularly heard in trade forums is that, jobs, tax receipts and national economic power are interrelated components of successful international value-added network services (IVANS) agreements.

Generally speaking, we can identify six types of IVANS: electronic mail services (E-mail), database access services, information processing services, packet switched services, gateway services (i.e, links between domestic networks and overseas networks), and private corporate networks. However, the problem of definition remains in where the line is drawn between basic services and value-added services. For example, many argue that ISDN services are value-added services, while in some countries, companies which offer packet-switched data communications services are considered VAN operators. Thus, while a number of administrations, notably the Americans and the Japanese, have sought to define what is and what is not a telecommunications service, it is perhaps not surprising to find that different telecoms authorities maintain different views of what to classify as a value-added telecommunications service.

In Asia, the distinction between value-added and basic services has become a focal trade issue. Asian governments are reluctant to allow foreign companies close access to the heart of their country's information industry.[24] It is for this reason that a number of regional governments have attempted to distinguish between telecommunications VAS (value-added services) and VANS (value-added *network* services). This is meant to distinguish between the use (by MNCs) of IVANS to provide their businesses with international services, as against the use of IVANS as a vehicle for investing in domestic network developments. Prominent examples of this 'fine line' are Japan and South Korea, where international resale of real-time voice services is prohibited, but where the provision of enhanced voice services (store and forward messaging and similar services) is acceptable (Tables 5.13 and 5.14).

State policy-makers across Asia thus find themselves caught in a dilemma: on the one hand, growth in the trade of value-added services is seen as a spur to economic growth. On the other, they remain well aware that foreign companies, especially American and Japanese ones, are regularly backed by their governments. This, and the pursuit of bilateral arrangements, tends to harm the credibility of free trade advocates.

In Asia, Japan dominates the VAN business because of the number of its operators, its huge market size, and its economic power. With large numbers of factories and offices across the region, Japanese companies account for a significant percentage of VANS subscribers outside Japan. It is partly their demand for services that is pushing governments to deregulate, both domestically and internationally. There are five countries in Asia (other

Table 5.13
VANS Revenues in the Asia Pacific Region

SERVICES	1990	1995(e)	% growth
Network information (All on-line electronic services)	1,704	4,000	137
Network processing (Computer reservation systems, EFT/EFTPOS, interbanking)	1,181	3,722	215
Enhanced communications (public e-mail, fax refilling, EDI, voicemail, store & forward telex)	667	1,465	119
Managed network services (data, voice, VPNs, facilities management)	187	859	359
Other VANS (videoconferencing, INs, teleport services, telemetry)	134	428	219
TOTAL VANS REVENUE	3873	10,474	170

Source: *Asia-Pacific Telecommunications*, March 1992, p.13 (based on a survey conducted by Systems Dynamics).

Table 5.14
International Services Across Asia

	IVANS	IPLCs	ISR
Hong Kong	yes	yes	no
China	no	no	no
Taiwan	yes	no	no
Malaysia	no	no	no
Singapore	yes	yes	no
Indonesia	yes	no	no
South Korea	yes	no	no
Thailand	yes	yes	no
Philippines	no	no	no

IPLCs: international private lines; ISR: international simple resale.

than Japan) with advanced VAN services: Hong Kong, Singapore, Indonesia, South Korea and Taiwan. (Malaysia is likely to join this list in the near future.)

In Hong Kong, VANS were first provided by companies such as GE Information Services (GEIS), Reuters, Tymnet and Infonet in the mid-1980s. By 1993, more than 30 licensed IVANs offered a variety of services linking Hong Kong with Japan, the UK and the US. Restrictions on the use of leased circuits have been significantly relaxed to allow voice value-added services and managed data network services to be provided, and to remove the requirement that bilateral arrangements be in place prior to these services being offered. Singapore similarly has more than 30 VAS providers in the market besides Singapore Telecom.[25]

In South Korea, international VAN services were liberalized in July 1991 as a result of bilateral government negotiations with the US. By the end of 1991, 24 private companies were registered as international VAN service providers offering international VAN services such as database access, data processing and electronic mail. The domestic market was also opened up for EDI and database services by foreign service providers who can own up to 50 per cent of domestic VAN providers. Major Korean companies have tied up with leading VANS providers: Hyundai Electronics with AT&T; Samsung Data Services with IBM; and Systems Technology Management (STM) with EDS. All three have registered as IVANS providers. At the start of 1994, regulations against foreign investment in the IVANS sector were completely eliminated.

The story is similar in Taiwan, where US trade pressure has been successful in slowly prising open the international value-added service market. The VAN market for domestic as well as international services was opened, subject to certain limitations, to foreign companies in March 1992. An advanced service, the virtual international private network (VIPNET), providing international data, fax and voice communications between domestic telephone users and overseas designated correspondents was introduced in October 1993, and further categories of services have been liberalized as the Taiwanese government negotiates for membership in the GATT/World Trade Organization.[26]

---

[1] Although no precise figures are available, GATT estimates that the coverage of GATS including transactions by foreign affiliates is probably closer to US$2000 billion.

[2] In the words of one participant, GATS 'fudged on telecoms', sending it to the international equivalent of a 'too hard committee'. See *Telenews Asia*, 27 January 1993, p.10. Some of the more liberalized telecoms countries such as Australia and the US had been pushing for an agreement in basic services but had to drop this plan in favour of pursuing bilateral negotiations. Basic telephone services are to be referred to a 'future negotiations' committee, which is projected to take between 18 months and four years to complete its recommendations. Significantly, audiovisual services were also left out (at the last moment) of a comprehensive services agreement, so as to placate the European countries, in particular, France. This gap could potentially have profound repercussions given the 'convergence' between the audiovisual industry(s) and telecommunications, for example, in terms of satellite broadcasting (distribution) and video-on-demand. These issues are discussed below. The French were worried about liberalizing their cultural industries given the strength and global dominance of Hollywood.

3   It should be noted that these figures are only 'guesstimates' since traffic over the PSTN is recorded as flows of traffic and broken down no further. These figures are conservative and are probably low, they do, however, correlate well with the tabulated figures in Table 5.2.

4   The universality of the change here is highlighted by using the US statistics as a reference point against which to measure the Asian countries.

5   As a result of Taiwan's policy, all telephone business between the two sides must be routed through a third party, usually Hong Kong, Japan or the US West Coast. Consequently, both sides have paid significantly more for the privilege of talking to each other. According to Taiwan's DGT, they 'lost' an additional US$53.8 million to third parties, between June 1989 and May 1994. China's MPT claims to have paid third parties more than US$20 million for the single year between June 1993 and June 1994. *China Daily*, 11 August 1994, p.4.

6   This has resulted in renewed calls for a relaxation of Taiwan's 'no direct communications' policy and the installation of direct communications links.

7   According to Aamoth (1994), who provides an interesting perspective on American accounting rates, recent data suggests that the 'growth in the so-called US settlements imbalance may have stopped and the size of the imbalance may be as much as 40% less than FCC statistics indicate. . . . The adoption of accounting rate benchmarks and the filing of progress reports by US carriers appears motivated primarily by trade and economic objectives .' (p.28)

8   An example of the former is BT's Global Network Service (GNS). The launch in Indonesia in cooperation with Indosat, completed the ASEAN loop of the service to Malaysia, the Philippines, Singapore and Thailand. Other countries in the Asian region using GNS include Australia, Japan, Hong Kong, South Korea and New Zealand.

9   As a result of CP's China success, its TelecomAsia partner, Nynex, has been using the Group's Hong Kong-listed subsidiary, Orient Telecom & Technology (OTT) to forge a 'quieter' entrance into the Chinese telecommunications market, see Chapter 6.

10 For an explanation of 'the relative failure of ISDN', see ITU (1994) p.48.

11 There is a further component to this equation: the accounting rate system, which provides the contractual mechanism for sharing revenues among two legally distinct entities jointly engaged in providing international services. Three distinct rates are used. The *accounting rate* is established between two carriers as the basis for international settlements. Each pair of carriers negotiate a rate per paid minute of traffic that they use to settle imbalances of traffic flow between them. The carrier sends the traffic and reimburses the carrier that is receiving the traffic, by a fixed proportion of the accounting rate — the *settlement rate*, usually 50 per cent of the accounting rate. By contrast, the *collection rate* is the amount that each carrier charges its own customers. Each national telephone company, typically sovereign (monopolistic) on its own territory, has been able to charge its own customers whatever rate it wanted for international messages.

12   The half-rate regime has thus encouraged cost sharing and broad multilateral ownership of facilities without jeopardising exclusive national rights. (See Staple, 1992.) The dominance of the PTTs in the international trade was reinforced by the sheer size of the assets they owned: satellites and submarine cables have significant scale economies. The bigger the facility and the more the costs are shared, the harder it may be for an alternative facility to be competitive.

12   Most carriers now charge the same prices for satellite and fibre circuits. The exceptions are Telstra and AT&T, which continue to levy a premium on fibre — up to 12.6 per cent for Telstra and up to 66 per cent for AT&T. HKTI has the lowest interregional tariffs by far. Its 64-kbps half-circuit charge for fibre, for example, is 40 per cent lower than the corresponding circuit from Singapore Telecom, 42 per cent lower than circuits from Telstra, and 70 per cent lower than from Japan's KDD.

13   The number of transponders counted has been for regional or sub-regional satellite systems. Thus, the Japanese satellites which, until early 1994, were not allowed under Japanese law to broadcast internationally, have not been included here. Also excluded are transponders on the Inmarsat satellites, satellites such as the Indian Insat, and Iran's Iransat — all of which would increase the total number of satellite transponders over the region. On the other hand, satellites launched or planned for launch by countries under consideration in this book, which will probably end up as primarily 'domsats', such as Measat, have been included because of the potential for overspill and because just by launching they are, as ThaiCom has found, having an influential impact on their neighbours. The numbers charted in the text are meant to be indicative, not exhaustive.

14   The history of the transoceanic cables has similarly been marked by speculation of over-capacity each time a new cable has been proposed, but by the time they are brought on-line, capacity is invariably utilized. See Elbert (1990) p.28.

15   For example, part of the deal with Rupert Murdock after Richard Li (second son of Hong Kong tycoon Li Ka-shing) sold STAR TV was that Li would not re-enter the satellite broadcast business. Instead Li, through his company Pacific Century, has been lobbying — quite successfully it appears — for deregulation in China, Hong Kong, Singapore and Taiwan to permit the use of transponders (from AsiaSat and other satellites) for direct intra-corporate communications. Such a by-pass of the national carrier is known as self-provisioning. AsiaSat is one-third owned by the Li family interests.

16   Satellites can be the most cost-effective way for developing countries to set up a domestic telecoms infrastructure fast, given the crushing capital cost of connecting remote rural areas with land-based systems. Furthermore, the first step that a country takes into the global communications marketplace will usually take place through satellite services. AT&T and NTT jointly invested in earth station facilities in Beijing to provide facilities which allowed their clients to integrate their China operations through private networks. More recently, Viet-

nam has been welcomed back into the international community through similar direct linkages provided by companies such as Telstra and France Telecom International.

17 For example, as an interim step ahead of major network expansion programmes, cellular operations have sprung up all across Asia. In Vietnam (three operators), Laos (two operators), Cambodia (four operators), Indonesia (five operators), India (eight operators) and even in Burma (one operator), cellular systems have been offered as a way around the state-supplied bottleneck. Cellular licences are held by satellite operators Binariang (MeaSat), TRI (Rimsat), Hutchison (AsiaSat) and Shinawatra (ThaiCom).

18 VSATs have traditionally been used to serve the financial and information exchange markets and, indeed, have been developed over the past few years particularly for data communications, providing network by-pass, quick response times and high availability. There is, however, great scope in the Asia Pacific region for small dish systems to provide rural telephony using Demand Assigned Multiple Access (DAMA). DAMA is a transmission and switching mechanism used to assign satellite circuits and earth station equipment on demand. It is based on a single channel per carrier (SCPC) technology where discrete bandwidths are allocated for each voice, data or fax circuit. Space segment usage is minimized by assigning the satellite circuit only when required. A DAMA network requires a central controller which can be located at any site, but is usually located at the main or hub earth station and thus is suitable for networks with a large number of remote stations with thin route applications.

19 The popularity of the service decreased when locally produced programmes about China were broadcast.

20 In 1994, Shinawatra was offering a rate of US$1.3 million for Palapa users to switch over until the contract with the Indonesian satellite expires. Shinawatra will also take over all costs including modifying equipment from other satellites to suit ThaiCom.

21 A telecommunications hub may be defined as a communications centre controlling telecommunications traffic in a geographical area with connections to other hubs in the network, thereby providing alternate routing to and from end destinations. See Langdale (1992) p.6.

22 Hong Kong's strong regional position can be attributed partially to the aggressive push Cable & Wireless made for this market by offering low tariffs in the 1980s.

23 Langdale notes while there is little public information available on the size of the regional telecommunication hub market in Asia, estimates suggest that the regional RTH market is growing at approximately 20 per cent per annum. (See Langdale (1992) p.8.

24 Equally, foreign companies are reluctant to share their key knowhow in ventures where they have only minority shareholdings.

25 In the IVAN market, Singapore Telecom is a prominent investor in Infonet and

there is a quasi-IVAN agreement with Japan. Other agreements are with AT&T and Hutchison.

26. In Malaysia, international VAN services are provided on a monopoly basis by STM. However, some limited liberalization is expected in the near future. VADS — a joint venture between Telekom, IBM and the National Equity Corporation (PNB) — operates data services to Hong Kong via satellite. In Thailand, foreign service providers like Japanese IVANS, Infonet, and GEIS are beginning to enter the market.

The Broadcast Media Markets in Asia

Peter Lovelock and Susan Schoenfeld

While the International Telecommunication Union tempers the more grandiose visions of telecommunications growth across Asia with the sombre realization that it will be well into the next century before the majority of the region's peoples have access to even basic telephone services, there are now very few places in Asia that have not succumbed to television. Indeed, to many it comes as a shock to realize that in a country like China, where there may now be up to 25 million households with a telephone, there are more than 200 million households owning *at least one* television (Table 6.1).

In fact, this case is more common than not across Asia. In the rural northeast of Thailand, in the rainforests of Sarawak, in the far eastern is-

Table 6.1
Demographics, 1993

Country	Population (mil)	Households (000s)	TV homes (000s)	% of HH	Population /TV HH
Hong Kong	5.8	1,637	1,596	97.5%	3.6
Singapore	3.1	752	720	95.7%	4.3
Malaysia	19.6	3,022	2,720	90.0%	7.2
S.Korea	44.6	–	10,806	–	4.1
Taiwan	20.9	5,313	5,233	98.5%	4.0
Thailand	59.2	12,255	10,049	82.0%	5.9
Indonesia	189.5	36,000	15,000	41.7%	12.6
China	1,200	300,000	230,000	76.7%	5.2
Vietnam	72.6	–	5,000	–	14.5
Philippines	64.8	12,211	6,960	57.0%	9.3
TOTAL	1,680.1	371,190	288,084	77.6%	5.8

Source: *Screen Digest*, April 1994.

lands of Indonesia, or on the Tibetan plateau, you are just as likely to find a satellite dish as a telephone, and more likely to find a television. To the state governments across Southeast and East Asia, this situation presents both problems and opportunities. A number of governments are hoping to fund the expansion of their telecommunications infrastructure by employing new technologies combining the transmission mechanism (the 'highway') of telephony with new pay television services, such as cable TV, video-on-demand, home shopping, and a host of other interactive and entertainment services. Although income distributions throughout Asia remain highly skewed, the rise of more prosperous, largely urban, communities is creating a local market for media products and services, and a commodification of local cultures, to which both governments and capital are responding, positively and negatively.

The emergence of this much vaunted 'Asian middle class' is also attracting the attention of international media and telecommunications companies. The industry journal, *Screen Digest* (April 1994) cites market predictions that by the year 2000 the number of affluent Asian households with incomes in excess of US$30,000 will have increased by 50 per cent to 51 million, and that Asian urban populations will rise by 37 per cent — a further 397 million people. Most Asian countries now have either cable or satellite television (see Table 6.2 and compare with 1994 year end data in Table 7.5) and a few even have promising pay TV networks; in consequence global programmers are now aligning themselves to exploit what they anticipate will be tremendous growth.

Table 6.2
Television Signals in Asia — by Wire and Wireless

Country	Satellite homes (000s)	Cable homes (000s)	% of total homes			% of TV homes	
			TV (%)	sat (%)	cable (%)	satellite (%)	cable (%)
Hong Kong	–	50	97.5	–	3.1	–	3.1
Singapore	ban	18	95.7	–	2.4	–	2.5
Malaysia	ban	–	90.0	–	–	–	–
S.Korea	500	6,000	–	–	–	4.6	55.5
Taiwan	300	2,922	98.5	5.6	38.0	5.7	38.6
Thailand	13	126	82.0	0.1	1.0	0.1	1.3
Indonesia	600	–	41.7	1.7	–	4.0	–
China	600	32,000	76.7	0.2	10.7	0.3	16.0
Vietnam	–	–	–	–	–	–	–
Philippines	30	250	57.0	0.2	1.1	0.4	3.6
TOTAL	2,043	41,366	77.6	0.6	11.1	0.7	14.4

Note: Cable TV includes MMDS (Microwave Multipoint Distribution System).
Source: *Screen Digest*, April 1994.

The purpose of this chapter is to explain the process of dynamic change that is now occurring in the media industries across Asia, and their relation (or not) to developments in the telecommunications industries in the region. The process of change is due to a combination of new technologies, increasing wealth, and new-found interest from both within and outside the region, and this chapter will highlight the developments which are the most influential in determining the changes in media markets in the region, the so-called 'drivers' of change.[1] In addition to the technology, also influential on the development of the regional media market are: the impact of multinational companies moving into Asia, government-to-business relationships, and regional business networks.

These driving influences have strong global features, but their impact upon the countries of Southeast and East Asia are as much regional, and indeed trans-regional, as they are national.Therefore the approach adopted in this chapter is thematic rather than country-by-country. However, to convey and highlight the characteristics of media markets in particular economies, the following sections also provide background details to one or two of the Asian markets under consideration:

1. Convergence — covers Hong Kong and Thailand
2. State Policies — covers Malaysia, Philippines and China
3. Regional Hubs — covers Singapore
4. Business Networks — covers Taiwan and South Korea

Convergence

Binary notation is the language of computers, a mathematical alphabet consisting of two digits, 0s and 1s, and any technology which uses this language to programme its operation is known as digital. Modern telecommunications networks are essentially a network of computers called telephone exchanges or switches. These exchanges 'digitalize' voice, data and image (for example, fax) into binary sequences of 'off' and 'on' electrical currents (or, if optical fibre is used, light) for high-speed transmission down wires and cables, or across microwave systems, to be electronically 'switched' through computerized exchanges.

The computers, telecommunications and television industries have very different histories. The computer industry has been part of the private sector, although early government support came from military and civilian contracts. (See Cohen and Noll, 1991.) The public switched telephone network traditionally has been in the hands of the state, or of a regulated private national monopoly, while television stations have been both state-run and privately owned. What digitalization has done is to bring these industries to a potential point of convergence, whereby they 'acquire like character independently'. In turn, this 'like character' has brought about the possibility of synergy, the 'combined or coordinated action',[2] of these industries.

Synergy has two dimensions in this context, a technological and a business dimension. The technological permits the different operations from these industries to be combined, for example, to transmit moving images as well as data and voice traffic down broadband optical fibre cables so that telecommunications and television services can be delivered simultaneously along a single transmission path. In this context, convergence of communications and information technologies really does constitute a communications revolution.

The business aspects of synergy are more problematic because of fundamental differences between these three industries. The computing industry, for example, requires substantial continuous high-risk investment in research and development, capital outlays which can only be recouped through the mass sales of hardware or software in a highly competitive market. Telecommunications networks buy technology, although some of the largest world carriers have equipment design and manufacturing businesses as well. Investment is lumpy, requiring large scale financial commitments to periodically build out a network, while smaller scale capital commitments are required to upgrade the network in terms of its quality and the scope of services as new software applications are made available. Traditionally, television investment, after the initial broadcasting or cable (narrowcasting) network has been built, falls mainly on programme production or purchases, which while financially continuous, requires significantly lower expenditures on a continuing basis than those required in the computing industry.

Multi-media

Beyond synergy comes synthesis, or 'putting together, making a whole out of parts'. The merging of computer, media and telecommunications services into multi-media products has raised considerable hopes, and hype, in these industries in the 1990s.[3] The convergence of information technologies has been a popular argument since the late 1970s. However, the argument has gained a new lease of life in the early 1990s. Examples of multi-media products already on the market include CD-ROM players, which provide the moving image, data and sound with varying degrees of interactivity that allows the user to manipulate or investigate the material; and virtual reality head-sets which allow users to create their own realities from a pre-programmed range of options. Video conferencing is another example of multi-media where real-time voice, image and data communications can create a virtual conference including participants around the globe.

However, beyond these potential products and services, business synergy among these three industries has been rather more difficult to achieve. One reason for this, in addition to the differences outlined above, is that convergence has been technological, with far less development on the product or service content side. The markets for the products and services cited above may be growing, but with the possible exception of CD-ROMs, they are nowhere near becoming mass markets. There is a distinction which remains between these industries despite convergence: the programme content and the transmission mechanism. Television, and the related broadcast and narrowcast industries, remains primarily concerned with content and not necessarily with the deployment or development of the 'pipe' or 'highway'. Thus, while cable TV is often cited as a prime candidate to provide multimedia services (television, telephone, fax, online information services, video-conferencing and so on) at this stage there is very little actual multimedia convergence, either in the services available or in consumer behaviour, beyond a common transmission of digitalized signals. The services are generated by and provided from different sources, from different operators, and are consumed by users on different equipment, be it the telephone, the fax machine, the computer or the television set.

So industry differences extend from their economic differences to product and service differences, and the problem of finding business synergies strong enough to override these is clearly proving difficult. One spectacular example was the on-off US$33 billion merger of the telecommunications interests of Bell Atlantic and the cable television interests of TCI during 1994. Examples abound of similar failures. For example, computer manufacturer IBM failed to make a success of its acquisition of the German telecommunications manufacturer Rohm, a specialist in PABX equipment, and later sold the company. Often the financial and business priorities of these sectors do not yet converge, and a management imbued with the business interests of one is unlikely to respond easily or flexibly to the business interests of the other. But still the efforts to find the synergies continue. For example, during 1994 AT&T put in a bid for the French computer manufacturer, Cie des Machines Bull, although the bid was subsequently withdrawn,[4] while TCI renewed its search for a tie-up with a telecommunications company, entering discussions with US Sprint. It seems that despite the difficulties in finding genuine business synergies, mergers, aquisitions and alliances are being driven by the need to develop strategies which position companies in markets which, because of their growth potential, cannot be ignored and yet have great uncertainties attached to them.

Southeast and East Asia provide an increasing number of cases to illustrate the trend of technical convergency and the hoped-for synergies; of business experimentation; and of commercial exploration of multi-media possibilities. One reason why the technologically deterministic arguments have been finding credence in various parts of Asia is the comparative immaturity of the networks — many countries are finding they can 'leapfrog' into new technologies, new systems integration and new network facilities. To this end, the telecommunications carriers have considerable financial strength; they also have expertise in switching traffic, a vital component of interactive multi-media technologies. Singapore Telecom, for example, showing its usual sense of strategic planning, has bought equity investments in Cambridge Cable Ltd and Anglia Cable Ltd, two cable TV companies in Britain. These investmennts allow ST to learn the business. However, the regulation of Singapore's planned cable TV industry remains uncertain, it is still not clear whether Singapore Telecom may be required to compete with Singapore Broadcasting. Whether in the UK or at home, the business experience will prove useful when the opportunity arises for Singapore Telecom International to invest in other Asian telecommunications and cable television industries.

And this is indicative of two further trends of converging business interests in the telecoms and media industries across Southeast and East Asia. The identification of national media industries as high-growth, lucrative sectors to enter, has induced telecommunications operators to become involved even where technological convergence may not yet be apparent, as in China. For the foreign operator, the potential convergence of telecoms and media technology has provided the potential for a 'back-door' entry into the highly protected Chinese telecommunications market by providing new generation technology and a range of services to administrative authorities other than the traditional telecommunications operator. Similarly in Thailand, the groups which have been induced to enter the new media environment have no relevant experience; they enter not with common technological characteristics, but common business interests. Each of these examples are examined below.

Technological Convergence: The 'Highway'

The issue of convergence arose in Hong Kong in 1985 when the Broadcasting Review Board raised the proposition, already familiar in Britain and the United States, that if the Hong Kong government wished to ensure more diversity in television programming available to the public and encourage competition in the PSTN upon the expiry of Hong Kong Telephone Company's (HKTC) exclusive franchise in 1995, then licensing a cable TV network and permitting it to carry telephone traffic would allow the establishment of a second network.

Bids were opened for a cable TV licence in 1989 with the promise of a domestic telecommunications licence in 1995. The timing was unfortunate because just as Hong Kong was adopting a duopoly model of competition in telecommunications, Britain — Hong Kong's model for this experiment — was about to abandon duopoly as inadequate. Further, the government paid little thought to the fact that, while technologies may converge, giving rise to operational synergies, the businesses of telecoms and television may not. The consortium awarded the franchise, Hong Kong Cable Communications Ltd (HKCC) was led by one of Hong Kong's largest property companies, Wharf Holdings, and included the giant US West telecommunications company, Coditel, a small Belgium cable company, the television production company Shaw Brothers (owned by Sir Run Run Shaw, the owner of TVB, Hong Kong's major television broadcaster), and another of Hong Kong's property developers, Sun Hung Kai. The sharing of management control of investment schedules and strategic planning proved too much, and when US West withdrew, apparently deciding there were more lucrative and lower-risk opportunities in Europe, HKCC collapsed without laying a cable. But HKCC was not helped by a government that seemed unsure of the rules it was to apply, especially with regard to restrictions on advertising, regulations governing the use of satellite dishes, and the scope and timing of a second telecommunications licence.

This confusion in government policy illustrates the problems convergency brings to regulators. Until computer, telecommunications and television technologies began rubbing shoulders it was normal, in Hong Kong as elsewhere, for government to regulate telecommunications directly through controls over prices and profits, to regulate television by supervising programming standards, and not to regulate the computer industry at all. The emergence of new businesses which straddle these industries creates problems of inter-departmental policy coordination, whereby departments may take very different views due to different objectives or due to departmental rivalries. (In Malaysia, awarding the cable TV franchise has become an administrative dispute between RTM, the traditional broadcast body, and the more powerful Ministry of Communications. Similarly in Thailand, TelecomAsia's announcement, in late 1993, that the two million telephone lines it was building in Bangkok would offer cable television services, initially met with objections from regulators. The issue was not one of technology or regulation, but rather how profits from the cable services would be divided among the various ministries.) Further complicating this are problems of definition to decide whether new businesses fall under existing regulatory requirements, and if they do not then whether they warrant regulation. Video-on-demand (VOD) is an example of this.

Case Study: Hong Kong

In Hong Kong, for example, the Economic Services Branch (ESB) regulates telecoms, the Recreation and Culture Branch (RCB) oversees television licenses and standards, and computer-related issues, such as data protection, are the responsibility of the Trade and Industry Branch. When AsiaSat, Asia's first privately owned satellite, was launched in 1990 by a consortium consisting of Cable & Wireless, Hutchison Whampoa and CITIC (China's official foreign investment arm), Hutchison applied for an uplinking licence to broadcast STAR TV (Satellite Television Asian Region) from Hong Kong across Asia. The Hong Kong government then got into interminable difficulties trying to decide whether granting an uplink licence broke the exclusive licence of Hongkong Telecom International (it was allowed), whether Hong Kong residents should be permitted to receive STAR TV's programming using SMATV (Satellite Master Access Television) satellite dishes (it was allowed), and if so whether STAR TV should come under the same regulations as the two terrestrial free-to-air broadcasters — ATV (Asia Television) and TVB (Television Broadcasts) — with regard to the Broadcasting Authority's supervision of programme standards (it does) and content (it does not). Also, whether STAR TV should be allowed to broadcast in Cantonese or carry advertising.

Cantonese programming restrictions were imposed to protect ATV and TVB, especially since STAR TV claimed their market lay outside Hong Kong, but STAR is received in Hong Kong and the restriction has since been partially lifted. In arguing for a licence, STAR TV claimed that its principal sources of revenue would come from the wholesale of satellite transponder (broadcasting) capacity to programme providers and rentals collected from hotels and other subscribers throughout the Asian footprint. Since advertising was not supposed to be a revenue mainstay of STAR TV, the government started from a position of protecting ATV's and TVB's advertising revenues by prohibiting STAR from carrying advertising.[5]

The availability in Hong Kong of STAR TV was cited by HKCC as the final straw. If this is to be taken at face value, it illustrates the sensitivity of business start-up plans in these industries. The government wished to promote cable TV and a second telecoms network, but got into problems over regulations and restrictions on the activities of the new entrant into the satellite TV business. Plans to promote a second telecoms network on the back of a cable TV network were not given much credibility by a report from the London-based consultants Booz, Allen & Hamilton (1988) which estimated that the costs saved in the building of a second telecoms network by combining it with cable TV amounted to just four per cent, and that 'the viability of the second telecommunications network operation will not depend noticeably on the fact of its common ownership with a cable television network.' (p.vi) At the same time, building the cable TV network separately from Hongkong Telecom's PSTN network would add 18 per cent to its cost and that there was 'no economic case for awarding the cable television franchise to a company other than HKT in order to facilitate the construction of a competitive second telecommunications network.' (p.vi) But the report did argue that separating the cable TV network from HKT's PSTN would encourage a separate focus upon the business itself. 'By combining management of the physical development of the network with the management of the required investment in programming and marketing, the risk of failure of the enterprise is minimized.' (p.xi) In other words, telecoms and television are very different types of business, and to make either successful they should be managed separately.

Following the collapse of HKCC in 1990, the government altered course. The new course was to separate the cable TV franchise, which again went to Wharf Holdings, but this time as a single operator, from new telecommunications licenses, of which three were approved, including one for Wharf to be operated as a discrete business. So TV and telecoms will still have an opportunity to converge, to 'acquire a like character independently', over a cable.

But in Hong Kong, as wherever else technological advance is allowed to be translated into market opportunity, the problems of regulating a moving target are endless. No sooner was the cable TV issue resolved when Hongkong Telecom proposed launching video-on-demand trials, demanding the right to operate commercially as soon as Wharf's three year exclusive franchise on cable TV ends in 1996. Is VOD a substitute for the video rental shop, or for cable TV? Should the content of VOD be subject to Broadcasting Authority controls, or is the private character of VOD distinct from the public character of cable TV? Is VOD a telecoms or a broadcast issue? The issues again become ongoing and their ramifications are as social and cultural as they are commercial, legal and technical. The latest policy does at least seem to recognize that to turn technological synergy into business synergy it is necessary to give room for each type of enterprise to first successfully establish its core business and expertise.

Business Convergence: A Commonality of Interest?

Elsewhere, however, industry developments are resulting from converging business interests, rather than the emerging synergies resulting from common technologies. Companies with little or no previous experience in mass media are suddenly converging on the industry, attracted by the high growth rates, and because Asian governments have been giving national priority to the related information technology and telecommunications infrastructure projects. Companies moving into these areas are thus often rewarded with beneficial state policies, such as periods of exclusivity.

Case Study: Thailand

The government and the military have dominated the media industries. Five channels are provided to a country with over 80 per cent penetration of the potential television market. The two principal operators have been the Department of Public Relations (DPR) and the Royal Thai Army. As a result, unlike Hong Kong, the companies entering the subscription television industry are relative newcomers. There are four main groups which are now seeking to provide services to the public.

Shinawatra Computers & Communications (SCC) originally distributed computers to the government. SCC controls International Broadcasting Co. (IBC),[6] the Bangkok-based cable television provider established in 1989, which now controls almost 90 per cent of the subscription television market. SCC also has an interest in ThaiCom, the national satellite, with an eight year monopoly on Thai media and telecommunications satellite traffic. In early 1994, the Communications Authority of Thailand, the government provider of international telecommunications services, approved a joint venture to share the exclusive right to uplink and downlink international programming with SCC. (SCC also has cellular and paging interests.)

The Charoen Pokphand (CP) Group is a Thai agribusiness conglomerate, producing animal feed and chickens, which has recently diversified into the media business. (See below.) In association with Nynex, CP holds an 85 per cent share in TelecomAsia. Other media interests include a share in the regional Apstar satellites, and media-related investments in China.

The Loxley Bangkok Co. is controlled by the powerful Lamsam family, which controls the Thai Farmer's Bank. The company's trading group owns part of the Thai Telephone & Telecommunications (TT&T) project which is building, transferring and operating one million phone lines in the Thai provinces. Loxley also established an uplink facility for STAR TV in northern Thailand.

> *Wattachuk*, a major publisher, bought the second Bangkok pay-TV operator, Siam Broadcasting and Communications, also known as ThaiSky. ThaiSky's more than 10,000 subscribers make up approximately 10 per cent of the subscription market.
>
> While none of these companies bring much in the way of broadcast expertise to their new holdings, all are looking to expand their media-related investments. CP, SCC, Loxley and Wattachuk are all among the companies bidding for the first privately held terrestrial broadcast license to be issued (1995) for Bangkok. In Thailand then, 'convergence' is not being driven by the long-standing media or telecommunications operators (the government and the military). Instead, convergence is being pushed forward by a few large trading houses which view both the media and telecommunications industries as lucrative sectors to be involved in.

Local Programming: Opportunity for Localization?

It is increasingly recognized that *national* media industries are influenced by *global* processes; it is less well-recognized that *regional*-level processes are also important. A central argument in this chapter is that to understand the Southeast and East Asian media industries it is necessary to recognize the processes working at all three geographical scales. In this section we look at the concept of the regional, or pan-Asian, media market, the strategies that have been employed by the media multinationals to distribute their product to the region, and the resulting focus upon local programming.[7] The following sections examine specifically how the states have reacted to this media 'invasion'.

In 1991, CNN International (CNNI), beaming down from the Indonesian satellite Palapa, and the BBC's World Service Television (WST), beaming down from Hong Kong-based AsiaSat1, went head-to-head in Asia. Previously, very few of the large Western media companies had focused on the Asian countries outside of Japan. There were a number of one-off operations, such as Rupert Murdoch's ownership of Hong Kong's *South China Morning Post*, and a few regional editions of global publications, such as *The Asian Wall Street Journal*, and the Hollywood majors were in evidence trying to protect their software from piracy. Asia was not generally considered a major media market. The arrival of the BBC and CNN signalled two things. First, that certain parts of the broadcast market had the capacity to globalize.

Second, that the providers of those services now saw Asia as a market worth competing for.

Retrospectively, one interesting aspect of that initial foray into Asia by the two international news services is that STAR TV (Satellite Television Asian Region), the first private regional satellite broadcast network which was just being created, was seen not as the main prize, but rather as the consolation.[8] Based on the US and European experiences, it was assumed by the operators that if any market in Asia could provide a profit, it would be Singapore. A pan-Asian service was simply too much of an unknown and most Asian nations were assumed to be far too poor to support such a service.

Ironically, having won the race into Singapore, CNN was forced to watch as HutchVision's STAR TV — and the BBC — grow from strength to strength across Asia. Where BBC WST was marketed as part of an innovative five-channel package, CNN was limited to systems with which it could make a specific deal (usually large hotel chains) and to nations that allowed their citizens to own private satellite dishes. There is a further limitation even here. A private citizen, buying a satellite receiver dish for several thousand American dollars, could only aim it at one satellite, so why aim it at the Palapa satellite to pick-up CNN when STAR was offering the BBC and four other channels.

CNN responded by negotiating with the American cable sports channel, ESPN, and the international arm of the Home Box Office (HBO) network, to join forces on Palapa and market themselves as a triumvirate.[9] They were also able to go searching for other partners in a bid to counter STAR TV, which they found in the Australian broadcaster, AusTV, and Hong Kong's TVBI (the international arms of local terrestrial broadcasters). The loosely formed group was nicknamed the 'gang of five' by the region's press. The problem for the group however was that they 'owned' no satellite, as HutchVision (STAR's parent) did through an exclusivity agreement with AsiaSat, and that the satellite they were on, Palapa, had an inferior footprint to that of AsiaSat. More importantly, given the loose coordination of the group, they had no way of marketing themselves jointly and they had no ground infrastructure for collecting revenues, either from individual subscribers or cable distributors.

CNN, independent of their partners, continued to arrange distribution agreements with cable operators in countries including Indonesia, Thailand, Taiwan and India. In Hong Kong in early 1993, they secured an agreement with the Wharf Cable company, despite an existing agreement between Wharf and STAR TV. The Wharf pay-TV system, begun in October 1993, will be available to 1.5 million homes in the territory by late 1995, the largest single franchise in the world. Moreover, after Rupert Murdoch acquired a majority stake in the STAR TV service in July 1993, his News

Corporation group indicated that they intended to extend their own BSkyB news service at the expense of the BBC so as to reduce costs. They later followed this up by leaving the BBC off the northern beam of the service altogether.[10] As a result, both the BBC and STAR TV are now, independently, adopting CNN's marketing strategy and attempting to negotiate on a country-by-country and network-by-network basis.

Many in the industry have now begun to reject a regional model of broadcasting, arguing in favour of local sensitivity, local culture and local tastes. This has been particularly the case with many of the region's advertisers who object to the notion of a pan-Asian market because the region's diverse languages, customs and government regulations make it necessary to tailor advertisements for each market in Asia. What needs to be remembered here is that the broadcast or television market is still run on an advertising (and hence, ratings) model. Where telecommunications services are driven by quantity — the more telephone and data traffic, the more revenue — television production is, by and large, paid for by advertising, and advertising is about distinction, not quantity: the more channels available, the less chance there is for an advertiser to reach a large and captive market. Thus, until a subscription infrastructure is in place, the satellite and cable television 'revolution' remains a questionable proposition, at least in terms of revenue.

By 1993, for example, HutchVision had sold less than 20 per cent of STAR TV's available advertising airtime — less that 50 minutes a day. STAR's free-to-air service relies almost exclusively on advertising revenue, the potential of which, assuming full utilization at 1992 prices, was $US872 million.[11] In 1992/93 they were believed to be receiving less than 10 per cent of that. (See *The Asian Wall Street Journal*, 8–9 October 1993, p.12.)

A second, equally important, issue for the multinational broadcast groups, is that while the global impact upon national media industries may be accepted, it is a myth that US and other foreign programming is always the most popular in Asia. Hollywood's market share and revenues have simply not developed on the scale that might have been expected in Southeast and East Asia. In fact, local programmes are more often than not the most popular (Table 6.3). In addition, regionally based programmes, particularly Chinese programmes produced in Hong Kong, are popular among the overseas Chinese community in the region (Tracey, 1988).

For STAR TV, this realization meant that where they initially claimed their strategy to be one of aiming at the pockets of well-heeled expatriates and English-speaking Asians, they subsequently adapted their approach, first to multi-linguistic programming and, later, to a country-specific distribution strategy.[12]

STAR TV executives identified three target markets: an Indian market, a Middle East market and a Mandarin-language (Chinese-language) mar-

Table 6.3
Share of Local/ Overseas Programmes in Top 20 in 1994 (for select weeks)

Country	Locally produced (%)	Overseas acquired (%)
Hong Kong	75	25
Indonesia	85	15
Malaysia	85	15
Philippines	95	5
Singapore	na	na
South Korea	90	10
Thailand (Bangkok)	90	10
Taiwan	100	0

Source: SRG Group.

ket. They also began to chase distribution arrangements with operators in each of the countries under their footprint. STAR TV's belief in subscription collection stems from the fact that no one has ever profitably run a satellite television service based solely on advertising. No programme suppliers will allow their best products to be shown on a free-to-air network because of the piracy problem, and advertisers are wary of promoting their goods on systems where they cannot assess the penetration or the demographics of their audience.

In 1992, a census commissioned by HutchVision showed that at least 3.75 million households around the region were receiving STAR TV broadcasts. A 1993 census showed that this number had increased to 11.3 million (Frank Small & Associates, *Homes Penetration Report*, February 1993). A further survey in late 1993 showed a household population of 42.7 million households (Table 6.4). Where the 1992 census suggested to advertisers that STAR TV had a presence in India, Israel and Taiwan (these three countries made up 90 per cent of STAR TV's market), the 1993 surveys indicated just how rapidly change was occurring across Asia — by November, China alone made up 70 per cent of the STAR TV market. The figures also helped to confirm that STAR TV was itself becoming a brand name regionally, acting as a further catalyst for entrants into the regional broadcast market. There is fairly wide acceptance that STAR TV's published penetration figures have been *underplayed* in an effort not to frighten governments that have yet to fully sanction the service.[13]

As a result of their unexpected multicultural success, further emphasis was given to the Asian culture of the network. The MTV station became very much an alternative to its American counterpart.[14] The focus was placed upon the Mandarin station, the network's strongest channel, and a new Hindi-language channel was added to the service's southern beam. Ironically, some of the most successful programming broadcast through Asia by the network has been Japanese-made but dubbed into Mandarin. It is an ironic develop-

Table 6.4
STAR TV Market Penetration, 1993–94

MARKET	Jan 1993	Oct 1993	% of total	Dec 1994	% of total
China	4,800,000	30,506,483	72.2	38,200,000	
Taiwan	1,980,140	2,376,433	5.6	(Greater China)	70.7
Hong Kong	304,809	330,827	0.8		
India	3,300,500	7,278,000	17.2	12,300,000	22.8
Indonesia	36,211	49,807	0.1	530,000	
Thailand	32,393	142,805	0.3	(ASEAN)	1.0
Philippines	137,141	187,431	0.4		
Pakistan	61,239	77,118	0.2	1,400,000	
Kuwait	12,780	31,210	0.07	(Pakistan and	2.6
United Arab Emirates	72,809	116,589	0.3	the Middle East)	
Israel	410,000	621,000	1.5		
Saudi Arabia	–	368,940	0.9		
South Korea	18,945	183,838	0.4	1,400,000 (other countries)	2.6
TOTAL	11,166,967	42,270,481	100.0	54,000,000	99.7

Source: Frank Small & Associates.

ment on two counts. First, because the Japanese have been so hesitant to put up their own regional service for fear of being seen as having imperialist intentions.[15] Secondly, STAR TV cannot be received legally in Japan.

The resulting trends towards localization within this framework of regional media developments are probably best illustrated by the emergence of the contemporary Hong Kong production industry, particularly given the current attractiveness of Chinese film libraries to the pan-Asian broadcast groups. Cantonese movies used to dominate the Hong Kong market in the 1950s, drawing on traditional Cantonese cultural forms such as opera, musicals and contemporary melodrama. Their popularity declined in the 1960s and early 1970s, when American movies consistently outgrossed locally produced Chinese films. By the 1980s however, the most popular film genres in Hong Kong were once again locally produced, in the Cantonese language. By 1989, 78 per cent of Hong Kong's cinema box-office was occupied by local films, with the remaining 22 per cent being made up by foreign, mostly US titles. (See *Variety*, 2 May 1990, p.28.)

What had evolved was neither 'foreign' nor 'indigenous': locally made movies that borrowed heavily, and successfully, from the Western action-adventure format. What emerged was a highly distinctive and economically viable cultural form in which the 'global and local are inseparable, in turn leading to the modernized reinvigoration of a culture that continues to be labelled and experienced as Cantonese' (Ang, 1991, p.7). Thus a strong domestic industry which sold successfully in Taiwan, China, Singapore — all countries with sizeable Chinese populations — as well as Thailand, Malaysia, Indonesia, Vietnam and the Philippines has developed.

For TVB, which produces over 85 per cent of its Chinese channel programming locally, this has meant a growing regional market. The recent developments in satellite broadcasting and pan-Asian distribution have made both the Hong Kong stations (ATV and TVB) very attractive media assets because of their programme libraries and their local programming expertise.Regionally, therefore, we are witnessing the emergence of serious rivals to Hollywood's dominance in the cinema, and as a programme provider for the upcoming proliferation of television channels.

This tendency is of considerable importance given the dynamic economic expansion in Southeast and East Asia. With many of the political barriers to trade in cultural goods between states in the region being removed, the significant growth in the Asian economies is seeing the audiovisual sectors develop rapidly and the beginnings of significant intraregional television, video and musical programming.

The strategy of the international programme distributors has gone full circle, from country-specific to pan-Asian to market-by-market, demonstrating the fluidity of the current situation in Asia.

State Policies

When introduced, there were several practical reasons for doubting the prospects of the STAR TV service: a lack of cable networks across the region to redistribute terrestrially the satellite signals, and the lack of individual satellite dishes to make a direct-to-home (DTH) market sustainable; a lack of any common language through Asia, other than perhaps English; and the wariness of the advertisers whose revenue would be required for any service that were to remain unencrypted, or free-to-air.

Moreover, most nations initially either banned or actively discouraged reception. However, by 1993, very few Asian nations still had no contact with the service. Even the more outspoken critics such as Malaysian Prime Minister Dr Mahathir allowed reception in hotels and embassies and, in Malaysia for example, there has been a growing illegal penetration in the north near the Thai border and along the more urbanized West Coast.The result was that the countries of Southeast and East Asia (also South Asia and the Middle East) began to react to regional broadcasting. In other words, just as STAR TV was forced to respond to the government regulations, so too the governments of Asia felt compelled to react to STAR TV and other satellite broadcasters.

In a number of cases, the initiation of the STAR TV service prompted what has since become known as the 'Singapore model'. In an attempt to counteract the attractiveness of the new satellite broadcasting (MTV, BBC and Chinese-language programming on the STAR TV service, CNN and others on the Palapa satellite) the Singapore government rapidly developed a nationally competitive system over which they could exert greater control — subscription television. (See Table 6.5 for the penetration of subscription

Table 6.5
Subscription (Pay-TV) Market (1993–94)

	Monthly rates (US$)	No. of subs (000s)	Annual revenue (US$mill)	By 1995
Hong Kong	26.40	90	28.5	1.5 million potential customers; largest single cable franchise in world.
Singapore	31.00	18	6.7	Government expects in excess of 150,000 subscriptions.
Malaysia	19.60	NA	NA	Both cable and TVRO revenue collection in place by 1995; size of subscription will depend largely on political considerations.
S.Korea	NA*	NA	NA	System announced in 1994; limitations upon foreign participation and content problematic for transborder operators.
Taiwan	22.00–24.00	2,922	806.5	Operators see this as most lucrative market by 1995 when cable licences awarded and implemented; penetration in excess of 60%.
Thailand	24.00	110	31.7	Further development expected; but little beyond metro-Bangkok; current MMDS system needs to be go to cable; perhaps 10% penetration.
China	1.00–1.50	32,000	480	Facilities will continue to grow exponentially, dependant upon the political climate.
Indonesia	NA	NA	NA	In many ways, could parallel China but slower, poorer, more political.
Philippines	5.00–14.00	250**	30	Some rationalization of cable systems expected by 1995. Revenue collection outside urban areas seen to be problematic.
TOTAL		35,390	1,383.4	

NA: in some cases these markets are TVRO (television receive only).
* the government has said the subscription rate will be set somewhere between $13–20.
** there are somewhere in the vicinity of 700 cable operators existent in the Philippines. Only a couple of these markets have been regulated to any degree.

television across Southeast and East Asia.) This way they can pre-view and control on a programme-by-programme basis material considered to be offensive or politically unacceptable without being seen as completely restrictive. In the words of Singapore's Minister for Information and the Arts:

> As satellite dishes get smaller with higher power satellites and clearer signal processing, it will become impossible to stop [alternative] television from being received. We must prepare for the change and go systematically into international broadcasting ourselves. The best defence is a good offense.
> (*Business Times*, 11 March 1992, p.3.)

Where this policy has often been portrayed in the press as defensive, using such terms as 'satellite ban', it is perhaps more correct to see such policies in the broader context of the managed state.

While Singapore has attempted to restrain the proliferation of satellite dishes, it has also felt forced to confront the issue. With its stated objective of being a world financial centre, financial institutions, both international and domestic, require access to real time news if they are to remain competitive. The contradiction between Singapore's economic objectives and its political stance towards foreign information and satellite reception was brought into stark relief during the 1991 Gulf War. The lack of access to real time news made it very difficult for companies to make critical financial decisions on a timely basis. Hence, the establishment of a licensing process for financial institutions and hotels to be able to receive satellite transmissions (primarily, at that time, CNN) and the use of CNN on delayed retransmission for one of Singapore's newly established CableVision channels. Since 1992, this model has been replicated, or planned, in South Korea, China and Malaysia.[16]

Indeed, since the start of the 1990s, all of the governments of Southeast and East Asia have responded in some fashion to the 'threat' of cross-border media broadcasting as Western media multinationals have moved in to establish a regional presence. This has resulted from a combination of three concerns: a desire to prevent, or at least monitor and control, foreign sources of information in particular news; a desire to promote national values, particularly where these clash with contemporary Western entertainment norms, such as sex and violence; and a desire to protect domestic programming and promote what is now seen as a key economic sector, the local media industry.

A Key Economic Sector

Because the mass media in most Asian countries has been owned or closely tied to government, the state has exercised much more influence over the course the broadcast media has taken than in the West. Where private stations have been allowed to open, in Indonesia and Malaysia for example, they are held by conglomerates whose owners are closely linked to the ruling party. In Malaysia this includes the New Straits Times, Utusan Melayu and STAR Publications. In Indonesia, the President's children and close associates maintain a disproportionate control of the media.[17]

In 1993, the Indonesian government ceded control of their satellite programme to the Satelindo company, 60 per cent owned by Bimangrah Telekomindo, 30 per cent owned by PT Telkom and 10 per cent by PT Indosat, Indonesia's international telephone company. Bimangrah's shareholders include Bambang Tihatmojo, President Suharto's third child, his associates in the Citra Binmantara corporation, and two businessmen who have handled investments for the Indonesian army's main pension fund. (See *The Asian Wall Street Journal*, 28 June 1993, pp.1,4.) Bambang also owns the RCTI television station.[18] About a year after RCTI commenced broadcasting in 1988, Sudwikatmono, a cousin of the President, opened his Surabaya Station (SCTV). Tutut, the President's eldest daughter, then started a morning educational broadcast (TPI) in December 1990, renting TVRI transmission facilities. The President's youngest son, Tommy, is also known to be planning a national sports network. Apart from the Presidential family, close Chinese associate Lim Sioe Liong's Salim Group, probably Indonesia's largest conglomerate, has started the Merdeka Citra Televisi Indonesia station in Semarang. (See *AsiaView*, November 1992, p.3.)

The reasons for these associations are to be found in the assessment by Asian governments that the media industries are high-growth and highly profitable industries. There is a temptation to exercise political favouritism and to share the financial benefits among close relatives and associates. On the one hand, we find many cases of government-linked companies, organizations or individuals running new media ventures. On the other hand, there is a local lobby to protect these key infant industries from foreign domination. Again political privileges are involved. Government assistance ranges from restricting foreign imported media (e.g., tariffs, quotas, delays on imports), or restrictions on foreign media companies (e.g., banning foreign firms, requirements to work with local firms in a minority equity position, or restricting areas of the media in which they can work), to providing subsidies to local firms. The Malaysian government, for example, instituted a cabinet ruling that its television stations could broadcast only eight hours of Chinese programming a week, in an effort to develop its national film industry.

Case Study: Malaysia

In Malaysia in 1993, the government-linked Renong Berhad group sold its media interests, including the leading newspaper publisher (the Fleet Group) and the most popular television station (TV3). This was allegedly done to end the government's direct involvement in business. In fact, Renong's media interests are now under the sway of a group known to support deputy-Prime Minister, Anwar Ibrahim. (See *Asiaweek*, 20 January 1993, p.43.) Furthermore, the country's second private television station, the proposed TV4, is controlled by Metropolitan TV (45%), a subsidiary of the Melewar Group, a conglomerate controlled by relatives of the sultan of Negeri Sembilan. Another 30 per cent of the equity is with the media group Utusan Melayu (a newspaper group with strong connections to UMNO), while the balance is with a consortium of six local companies: Berjaya Group, Nanyang Press, Kemas Runding, Rediffusion and Telekom Malaysia.[19]

Previously, there were only two television networks in Malaysia, the state-owned Radio Television Malaysia (RTM), with two channels, and the quasi-private, Sistem Televisyen Malaysia (TV3).[20] In early 1991, RTM began the process of introducing a pay-TV news channel aimed at urban and expatriate audiences, and by late1994 Malaysia was in the process of installing a wider system of subscription television, carrying foreign news and entertainment programmes, called Subscription News Service (SNS).

The parallels with Singapore in these developments are quite significant, for despite hectic activity the government has stood firm on its decision not to allow satellite broadcasts from any other operators in the region.[21] Moreover, as appears to be the intention with the Singapore Broadcasting Corporation, in the longer-term the buy-out of Renong-Berhad is seen as providing the basis for what could one-day be one of the largest privately owned media groups in Southeast Asia. Plans have already been announced by Renong to establish joint broadcasting and print ventures with media companies in Thailand, Indonesia and other ASEAN countries.

Case Study: the Philippines

In contrast to Malaysia, Philippine mass communications has been owned by large business groups for generations. Oligarchies such as those of Soriano, Lopez and Elizalde controlled newspapers, magazines and radio and television stations along with holdings in sugar, utilities, beer, and pharmaceuticals. After Marcos obliterated this media structure in 1972, his friends constructed and controlled new media combines, along with their widespead interests in electronics, telecommunications, shipping and the coconut industry. From 1973 to 1986 only four individuals controlled the country's media infrastructure — Roberto Benedicto, Brig. Gen. Hans Menzi, Benjamin Romualdez, Kerima Polotan Tuvera — all associates of President Marcos.

Since the 1986 revolution, a number of the old entrepreneurial groups have re-emerged, notably Roces, Lopez and Yap. In 1988, a bill was introduced in the Senate banning monopolization of broadcast media because of the fear that media conglomerates would be formed anew. However, the concentration of the media in a relatively small group of families and individuals proceeds unhampered, despite the legislative limits (Table 6.6).

Cross-media restrictions introduced in 1993 do, however, seem to have had an impact. The Lopez family, for example, having set up a holding company (Benpress) to participate in a fixed wire telecommunica-

Table 6.6
Ownership of Philippine Broadcast Media

OWNER & BUSINESS	TELEVISION NETWORKS	OTHER MEDIA
Lopez family — public utility (MERALCO), sugar, finance	ABS-CBN 2; 13 relay stations; 4 originating stations and 10 affiliate stations; SkyCable	DZMM-AM, DWRR-FM and 4 affiliate radio stations; Manila Chronicle
Roces family	ABC-5; 5 stations	Kool 106-FM(DWET)
Roberto Benedicto — sugar, agribusiness, manufacturing	RPN-9; IBC-13; BBC-2; 7 originating stations	BBC(6 stations), RPN(14 stations)
Jimenez Group	GMA-7; 1 originating station; 11 affiliate and 10 relay stations	GMA, DWLS-FM, DWRT-FM, DZBB-AM, DYSS-AM;
Canoy family	Channel 31(UHF); provincial TV stations	Radio Mindanao, 17 AM stations, 7FM stations, 10 affiliate stations
Catholic Welfare Organisation	DYAF-TV (affiliate of GMA-7)	21 radio stations, 18 AM and 3 FM stations
Pacquing family	Channel 21(UHF); station in Davao	A series of radio stations
Arcadio Carandang	Channel 23(UHF)	6 radio stations
Ramon Jacinto	Channel 29(UHF)	radio stations

Source: IBON, *Facts & Figures*, July 1992.

tions project, announced that it would have to sell its 30 per cent share in the *Manila Chronicle*. On the other hand, the ban will not affect Antonio Conjuangco, president of the Philippine Long Distance Telephone Company, who bought the controlling interest in the *Manila Chronicle* in 1992 from Lopez, as Conjuangco has no investments in broadcast networks.

In the Philippines, cable television began as a community antenna TV (CATV) system in 1969 in Baguio City. Due to the mountain ranges of the summer capital of the Philippines, Baguio residents could not receive the TV signals from Manila. Thereafter cable TV was largely controlled by the media monopolies, close Marcos associates and, in particular, Roberto Benedicto. In 1987, President Aquino demonopolized the cable industry and since then over 90 cable operators have emerged. Comparatively, however, the Philippines cable industry is still in its infancy. Local operators have remained essentially local, thriving mainly in areas not reached by free TV. However, there are some emerging exceptions.

The Philippines is the only country in Southeast Asia to have inherited the American concept of competitive commercial broadcasting.[22] There are six VHF television stations: ABC, ABS-CBN, GMA, IBC-Channel 13, People's TV, Radio Philippines Network (RPN) and World TV. There are also four UHF stations: channels 21, 23, 29 and 31. The mass market is dominated by two of the VHF stations — ABS-CBN and GMA, with ABS-CBN providing broadcast coverage to 71 per cent of the Philippines. This figure can be expected to rise with the expansion of affiliate networks.

Most of the six major networks have, over the last few years, shifted to local programming, a trend that is expected to continue. Only channels 5 and 9 have continued to carry a heavier load of foreign shows in their programme mix, and even channel 5, which caters primarily to an upscale market with 51 per cent of its programming allocated to foreign shows, has been adding more local programmes in an effort to increase both ratings and advertising.[23]

Promotion of National Values

The distinction made above, however, should not be read simplistically. Governments take protectionist stances for both cultural and economic reasons. The media is a powerful force in shaping culture and many governments want to use it to help unify disparate ethnic and regional groupings in their countries. In Indonesia, the mass media have been explicitly viewed as the single most important mechanism for promulgating a vision of national unity, capable of overcoming the diversity of this multiethnic state (Hamilton, 1992; Isaac,1992). In Malaysia, with its commitment to ethnic Malay nationalism and Islam as the state religion, the *nature* of television programming is equally important.

Thailand, likewise, sees the mass media as a primary mechanism for national development and ideological conformity. Each of the five capital-city television stations is owned by one department or another of the government. While stations and air-time are subsequently leased to private enterprise, overall control lies squarely with the state. One should note, however, relations between competing ministries *within* the government are often very intense and this will often translate into a constantly shifting ground of policy and practice as the result of political changes at the top of the administration or bureaucracy. Across Asia over the last several years this has often meant struggles between ministries of broadcast (or communications) and ministries of telecommunications for the rights to develop national cable (or satellite) television, or video-on-demand projects.

Restriction of Undesirable Western Influences

Somewhat relatedly, Asian governments often want to minimize what they see as undesirable aspects of Western media, particularly what is seen as an undue emphasis on sex and violence. Many governments are also sensitive to criticisms made in the media.

Case Study: China

In 1993, television penetration was approximately 81 per cent of the country, meaning that with some 900 million viewers and well over 200 million television sets in use, China was already ranked number two of the top sixty television markets based on total television sets in use, and number one based on total population.

In September 1993, the State Council issued regulations to control the installation and use of TV satellite dishes. Under the regulations, individuals or organizations are required to obtain government approval before being allowed to have satellite reception equipment installed. Those in possession of satellite dishes are required to register their equipment with the authorities, while those wishing to produce, import or install the dishes are required to obtain government-issued licences. Furthermore, sales and installation companies became required to be both registered and capitalized to a minimum of one million yuan.[24]

However, the distinction between the urban and the rural in China is a crucial one, and ministry officials made little secret of the fact that satellite regulations were to have no effect in rural areas. The satellite dish restriction is strictly an urban policy. Across rural China satellite broadcasting is a necessity for extending government communications. With mountainous topography making up almost 70 per cent of China's landmass, microwave conveyance of television programmes was effectively restricted to less than 40 per cent of the country prior to the introduction of satellite distribution in 1985.[25] Thus, since its launch in 1989, AsiaSat-1, one-third owned by CITIC, has been used to deliver China Central Television (CCTV) across the nation. Ironically, AsiaSat is also the satellite from which News Corporation's STAR TV is beamed down to the region.

When the satellite dish regulations were introduced there was much speculation in the international press that they were directed at STAR TV in particular, and at other foreign satellite broadcasters in general; in other words, that the regulations were the act of a reactionary and insular Chinese leadership. While there probably was some degree of truth in suggestions that the Chinese bureaucracy are not altogether conversant, or confident, with these new technologies, to interpret policy statements as being dictated by fear and xenophobia is perhaps to miss the objectives to which the Chinese leadership is moving.

Officials in the various ministries, such as the Ministry of Radio, Film and Television (MRFT) are quite categoric: China is not going to be a TVRO (Television Reception Only) or DBS (Direct Broadcast Satellite) market, *except* through the rural regions, which are particularly poor markets. (In other words, Rupert Murdoch is being overly optimistic if he thinks he can get $US20 per month out of the average Chinese peasant.) For those wishing to broadcast by satellite to China's rural regions, the demand for anything other than Chinese-language programming is extremely low. Across China's urban areas there is very little English-language fluency. In the rural areas there is next to none.

As for News Corporation's claims to have more than 30 million Chinese households tuning into STAR TV's broadcasts, the truth is far from that. Most of those homes are linked up via urban cable networks, and very little of the 4-channel (previously 5-channel) satellite selection has been getting through, most is simply edited out. Most Chinese with access to STAR TV get to see the Chinese Channel, and some of the sports channel. No 'STAR Plus', no channel V, and certainly no BBC.

Cable television is not even particularly new in China. It emerged in the 1960s and 1970s when manufacturing enterprises set up private systems as part of the benefit packages for employees (Table 6.7). In 1980, a

The Broadcast Media Markets in Asia   171

Table 6.7
Cable Licences Awarded in China, 1993

Province/Region	No. of licences	Urban networks	Private networks
Beijing	7	2	5
Tianjin	8	2	6
Hebei	11	5	6
Shanxi	65	6	59
Inner Mongolia	5	1	4
Liaoning	26	13	13
Jilin	10	1	9
Heilongjiang	77	1	76
Shanghai	7	2	5
Jiangsu	30	8	22
Zhejiang	13	11	2
Anhui	19	12	7
Jiangxi	49	16	33
Fujian	4	4	0
Shandong	38	8	30
Henan	33	10	23
Hubei	47	22	25
Hunan	1	1	0
Guangdong	60	60	0
Guanxi	12	9	3
Sichuan	28	20	8
Yunnan	11	8	3
Shaanxi	46	2	44
Gansu	21	1	20
Qinghai	4	0	4
Ningxia	1	0	1
Xinjiang	13	8	5
TOTAL	646	233	413

Source: Ministry of Radio, Film & Television.

draft proposal was submitted to the State Council requesting that all new buildings include an accommodation for MATV (Master Access Television). This proposal was accepted in order to improve the signals of over-the-air TV and the government has since undertaken a large investment to provide for MATV throughout China. In 1988, the government allowed operators to connect all of the MATV buildings with coaxial cable and a headend, thereby creating what is now referred to as an 'urban network'. These networks are typically 300 MHz and consist of 12 to 13 channels. Ownership of the urban networks is governed by the state and only one network is allowed in any city.

What Table 6.7 also demonstrates is the speed with which the Chinese government is attempting to roll out a cable-TV infrastructure, either broadband or MMDS. In the first nine months of 1993 alone, 646 cable-TV network licences were awarded. By the end of 1993, this was up to 661 licences. There are 2900 cities in China, all are targeted for a subscription television network.

Given the sizeable populations, many of the urban networks now coming on-line have the potential to be huge. Several of these systems are already noteworthy. In Shanghai, the network, wired with optical fibre, is in the process of being rolled out to a potential four million subscribers. By mid-1994, the system had already more than one million subscribers. The Sichuan Allday TV Development Corporation, a joint venture between Wharf Cable (of Hong Kong) and the local Sichuan Province Cable TV Enterprise, is using microwave transmission to provide television to the 22 million people in Chengdu and twenty neighbouring districts. The service has been launched with four subscription channels.

The urban networks charge installation fees of US$20 to $30 and monthly fees of US$1 to $1.50 (the lowest in Asia). Due to the low cost and lack of alternative entertainment, these systems are achieving high penetration levels. If you can receive STAR TV's Chinese Channel along with at least 12 other offerings, all in Chinese, for 10 yuan a month, who would want to buy a satellite dish?

There are, however, problems. These urban networks currently tend to comprise of tens, hundreds, or even thousands of small CATV networks: 400 or so families in a residential block. These networks are not yet interconnected and the technology which has so far been deployed is not consistent. The main problem is there is no way to turn off individual connections, and hence there is no way to enforce subscription. Thus, the cable service is offered to an entire block (or small community) or not at all. In Beijing, for example, people in one apartment block but with two apartments can receive cable TV programmes while paying subscription for one apartment and not the other. Chinese engineers in the institutes are thus working on developing decoders.

The sticking point is the manufacture of integrated circuits. Until now, the bulk of integrated circuits used in China have come from Hong Kong, South Korea and Japan. But the Chinese believe if they continue to import this basic electronic equipment they will miss out on a crucial step in the development of a manufacturing and high technology base and that will set them at a permanent disadvantage as the economy opens up. On the other hand, with such a huge domestic economy, there exists a potential to create their own demand. Hence, people are encouraged to want a television, and hence the television production base increases rapidly. Hence

> also, the requirement for AT&T and Motorola to set up integrated circuit production centres in China — trade access agreements in telecommunications that have come with very specific technology transfer demands. The point worth recognizing then, is that there is an economic (or state-strategic) imperative behind the satellite dish ban and the programming blocks.
>
> There is a further strategic aspect to this planning. By deploying a broadband network initially, the government will be able to provide both television and telephone services at the same time. For a country that is able to provide telephone services to roughly 2 per cent of the population, this is a significant consideration. Officiating at a ceremony, the governor of Hebei was quoted by the *South China Morning Post*, 12 August 1993, p.B14, as suggesting that the old wisdom that China's provinces would get rich first by building roads, then putting in telephones, no longer applied. Now, he said, the reverse was true. A further attraction to this strategy is the possibility of using cable-TV revenues to fund the telecommunications development programme.

What must be emphasized then is that the role of government in regulating the media industries across Asia is in a state of flux. This is partly because of trends towards local liberalization and partly because of the globalizing influences of multinational companies. To return to Singapore, the government was forced by advances in technology and by its desire to become a global and regional information hub to begin media liberalization. They have prohibited the use of satellite dishes but at the same time they have begun to privatise the broadcast sector and to introduce foreign material such as CNN through cable distribution. Other Asian nations are devising their own strategies for coping with the new technologies (Table 6.8).

The impact of new technologies and the internationalization of production have facilitated the internationalization of the media industry. While it is commonly argued that these forces will restrict the ability of Asian governments to regulate their media industries, this argument is only partly true. It is simplistic to assume that all Asian countries are totally defensive about their media industries. It is also difficult to separate out cultural from economic reasons for government regulation, but there is now a growing interest among Asian countries in attracting the economic benefits a vibrant media sector brings. However, this presents countries with a dilemma: on the one hand, they want to attract foreign investment and develop export markets for these industries in neighbouring Asian countries; on the other hand, they want to continue to strictly regulate content within their own media industries. These cultural and economic tensions are adding to the

Table 6.8
Regulatory Status of Satellite and Cable Systems

Country	Individual Dish Ownership	Hotel and Apartment Cable Systems	Commercial Cable Systems
Hong Kong	legal	legal	legal
Singapore	not legal (except financial institutions and hotels)	not legal	legal
Malaysia	not legal (except VIPs)	not legal	being set up
S. Korea	not legal (but common)	not legal (except govt and research)	trials underway
Taiwan	legal	not legal (common)	'semi legal'
Thailand	legal (licence required)	not legal	legal
Indonesia	legal	legal	to be legalised
China	requires a licence	legal	legal
Vietnam	–	–	–
Philippines	legal	legal	legal

dynamics of regional and sub-regional media and information hubs which are beginning to emerge.

Regional Hubs

> Our aim is to see Singapore develop into a regional broadcasting hub providing a base, or a one-stop service centre, for firms involved in up-linking, production, packaging and distribution of programmes for the Asia Pacific region . . . in five to ten years the Republic [will] be wired for multiple TV channels to provide a wide range of programme choices. . . . [this will] make Singapore truly an information hub with multiple local and foreign news channels, educational programmes for distance learning and interactive information services, apart from local and foreign entertainment programmes.
>
> Singapore CableVision (*The Straits Times*, 7 March 1992)

This issue of a regional hub in the Asian media industries has gained much recent attention because many now expect that there will be a dramatic increase in the number of television channels available. In Japan, with the three satellite networks and over 47,000 local cable stations, broadcasters are struggling to fill the airwaves. The three satellites alone have increased daily air-time from 140 to 230 hours. In Taiwan, regulators propose awarding up to 260 cable television licences, and there are currently in excess of 250 operators providing, on average, 40 to 60 channels per package (see below). The Wharf Cable network in Hong Kong is already in the process of increasing the number of channels on its service, up from eight originally to twenty by early 1995, and STAR TV talks of offering 'bouquets of nationally-focussed channel selections' with the launch of the AsiaSat2 satellite in 1995. In China, State Council plans call for all 2,900 of the nation's cities to be wired up to provide cable TV services, and already the services available — providing on average 12 channels — are having great difficulty filling their programming schedules.

What is foreseen, therefore, is the spawning of regional production centres as local redistribution systems (cable and broadcast) emerge, and pan-Asian satellite and distribution systems will continue to grow. As one or two regional centres (mini-Hollywoods) become dominant they will divert resources and creativity away from other national centres. So there is something of a race going on to tie-down distribution networks and become regional production hubs during this transition period.

Hong Kong has the most export-oriented media industries in Asia, while functioning as a regional centre of expertise for a range of post-production work. However, other Asian capitals — Singapore, Taipei, Kuala Lumpur and Bangkok — have been looking to capitalize on the uncertainty (particularly within the media industries) surrounding Hong Kong's transition to Chinese sovereignty in 1997. Ironically, it is precisely Hong Kong's integration into the booming southern China economy that has led many multinational advertisers to view the city as a headquarters for expansion into China, and potentially as the RHQ for a 'Greater China' and further out to Southeast and East Asia.

A more immediate problem for Hong Kong has been that until 1995 legislation has not permitted foreign companies to establish satellite uplinks. In a bid to encourage the international community, the government has announced that it intends to review and redraft the policy. Without a coordinated governmental approach such as exists in Singapore, however, this has presented problems. The Hong Kong government's Recreation and Culture Branch (RCB) has the job of developing and implementing broadcasting policy, licensing terrestrial and, as of 1993, subscription television, but the RCB has yet to develop a coherent broadcasting framework for the territory. In February 1994 it announced yet another consultative review in an attempt to find common ground for all interests. The Broadcasting Authority (BA) and the Television and Entertainment Licensing Authority (TELA) review RCB policy and monitor programme standards. Both the BA and

TELA are under the auspices of the RCB. The problem then becomes that free-to-air, cable and satellite broadcasters are all subject to different licensing criteria, codes of practices and licence fees. And finally, any major legislation which straddles the 1997 transition date requires ratification by the Chinese authorities as part of the ongoing negotiation process of the Joint Liaison Group (JLG).

This complex system remains opaque and has been cited by international networks such as Asian Business News and the Discovery channel as one reason for uplinking from Singapore rather than Hong Kong. Indeed, ESPN Asia, the American-originated sports channel, initially decided to uplink from Malaysia using Malaysia Telekom facilities to avoid these bureaucratic difficulties.[26] However, ways around the regulations can be found, with the Hong Kong government deciding that the transmission of encrypted signals through Hongkong Telecom by foreign companies is permissible.[27] CNN International, for example, located its regional production facilities in Hong Kong, rejecting Tokyo and Singapore. Hong Kong was chosen for its central location and the quality of the Hongkong Telecom uplinking facilities.[28]

Singapore's government has identified media as a 'second-generation services industry objective' (Langdale, 1992, p.13). This refers to the long-term objective to establish a regional information and services hub. In the first-generation of this strategy, Singapore attracted large multinational telecommunications companies and information providers, such as Reuters, and the banks and financial institutions, the belief being that the media industries will naturally follow in the next generation, because: (1) the infrastructure will be in place; (2) the interconnections between telecommunications and the media are becoming increasingly significant; and (3) agglomeration economies will develop, further increasing Singapore's attractiveness.

To this end, the Creative Services Strategic Business Unit (SBU) of the Economic Development Board (EDB) has been established to develop initiatives to attract foreign interests and nurture local industry. The first target has, of course, been Hong Kong. Attracting Reuters to relocate a part of their regional operations away from Hong Kong was therefore a major coup, as were the Asian Business News (ABN) and Home Box Office (HBO) decisions to establish their regional headquarters in Singapore.

Indeed, the establishment of regional administrative headquarters (RHQs) has been a development of growing importance in the media industries across Asia. This is because RHQs play a dual role: they serve to regionalize (or localize) operations — and in the cultural industries, this has obvious significance; and for the countries involved, multinational operations introduce technologies and other business influences operating at the international level. The relation between the establishment of regional head-

quarters and the emergence of regional hubs can be summarized into three related issues:
1. Where RHQs are established in a given region. Regional headquarters provide jobs, training and an obvious stimulus to the local economy.
2. The related companies that set-up to service regional offices. In the case of the broadcast media, these include not only the required production facilities and creative services, but all of the associated ancillary business services: consultants, accountants, lawyers, etc.
3. The agglomeration economies that establish an area as a regional production hub.

Cities with excellent telecommunications services are particularly attractive to the broadcast media, not the least because there is likely to be a concentration of skilled labour that can be used in a variety of broadcasting and telecommunications related industries because of the emerging interaction in broadcasting and telecommunications technologies. Thus, the convergence of business operations is an important consideration here. It is in this regard that one should view the current battle between Singapore and Hong Kong to be the regional hub in such sectors as satellite uplinking, programme distribution and audiovisual post-production.

Case Study: Singapore

The Singapore government operates three free-to-air television stations — 5, 8 and 12 — through the Singapore Broadcasting Corporation (SBC). Each of the SBC channels carries advertising and limits foreign programming to less than 40 per cent of broadcasting time.[29] Half of all foreign programming is from the United States, with 90 per cent of the foreign programming available on SBC 5. Most Singaporeans can also receive the three free-to-air Malaysian stations (the two RTM stations and TV3).

In 1992, the introduction of three cable television channels broadcast on a MMDS system by the government-owned Singapore CableVision (SCV) was permitted. The channels offered by SCV include VarietyVision, a Mandarin-language entertainment channel and MovieVision, with English programming provided by Home Box Office (HBO) Asia. There is also a 24-hour news channel, Newsvision, which carries CNN programming for approximately 22 hours and fills the remaining time out with ITN (Britain), NHK (Japan) and local programming.

The American media company Home Box Office (HBO) established its

> Asian RHQ — HBO Asia — in Singapore's NewTech Park in late 1992. The success of the operation is perceived important to both HBO's global strategy and Singapore's regional ambitions.[30] The HBO facility — a 20,000 square foot production centre — will be used for programme origination and the media company hopes to eventually handle all regional scheduling, editing, production, marketing and finance through the facility (*Post & Broadcast News*, 3 August 1992). For Singapore, the facility is of strategic importance since HBO Asia could act as 'a sort of subcontractor for other television ventures, offering office space, satellite uplink facilities and other services' (*The Asian Wall Street Journal*, 25 February 1993). The Economic Development Board (EDB) is also well-aware that other global and regional players will judge Singapore's regional pretensions by the success or otherwise of the HBO Asia venture.
>
> It remains the case that while the Trade Development Board has had some limited success with the EDB in encouraging foreign producers to set up production facilities in Singapore by promoting the provision of technical support services, logistics and business management expertise, there is no Singaporean feature film or production industry to speak of. Certainly there is nothing to compare to the Hong Kong production industry, or even to the small, but increasingly successful, industry in Taiwan. Moreover, many of the film and video support companies and much of the technical skill currently available in Singapore have been drawn from the Hong Kong and Australian film industries (*Business Times* [Singapore], 8 March 1991).

A second aspect the 'hubbing' argument which is often ignored is that regional cities also act as creative hubs. The role of cultural industries is becoming increasingly important to the vitality of cities, as centres for creativity and culture. The growing role of service industries in the international economy, particularly 'culturally intensive' industries such as media, design and advertising is leading to creativity having a growing role in economic growth. The vitality of a city as a regional media hub will therefore depend partly upon its ability to be a creative centre.

One of the problems international and regional producers have with Singapore is that while they can set up easily enough, they find that there is no ancillary support to produce a feature film; they tend to have to fly a lot of people and equipment in. Indicatively, while there are 20 commercials companies in Singapore and five post-production houses — including the largest such facility in the region — there are *no* programming companies other than the SBC's drama unit. The strategy of the Singaporean government is to develop the media industry infrastructure, believing that if the necessary production centres are in place, then the creative side of the industry will naturally develop around these facilities.

A further aspect then of the changing media dynamics in the region is that firms that have entered Southeast and East Asian media industries need to be considered not just as individual competitors, but in terms of the networks of relationships among firms in the industry, and as a part of the broader business networks in the domestic economy; and finally as an integrated part of regional and international industry development.

It is not only the multinationals that are working to gain a strategic initiative. National governments also, find it in their interest to form strategic alliances or encourage joint ventures. It is in this context that the governments of Hong Kong and Singapore have attempted to establish themselves as 'information hubs'. Equally, a number of second-tier information states, such as Thailand, Malaysia and Taiwan have attempted to position themselves as sub-regional hubs (e.g., serving Indochina or Greater China), or have targeted specific sectors of the media for development (such as post-production or publishing). This has been initiated in an effort to avoid becoming overly dependent on emerging regional centres. Similarly, almost every country in the region has plans to launch a national satellite. (See Chapter 5.)

There is a constant tension between *government objectives* (and the strategies they employ), *multinational* aspirations to develop pan-Asian marketing strategies, and *domestic business networks*. Significantly, it is the relationships between business networks and state governments in various Southeast and East Asian countries which are playing a determining role in regional media developments.

Business Networks

There is a unique relationship emerging between domestic developments and regional developments and, to some extent, these can be mapped out through the business networks involved. In Korea, for example, a combination of government intervention and industry-specific market growth has led to the emergence of a relatively concentrated group of dominant corporations, the chaebol. Similarly, in Taiwan a few large diverse groups have in recent times moved from positions of manufacturing dominance into the service economy. In Taiwan, as in much of the Chinese world, these corporations are usually overseen or initiated by a single Chinese family. In both cases, as elsewhere in the region, these corporations have used their dominant domestic position to expand regionally, capitalizing not only on their domestic economic strength, but also on the existence of well-established business networks. In the case of the chaebol this has resulted from their success at establishing regional manufacturing and electronics distribution networks. In the case of the Chinese, it is often the result of the overseas Chinese connections. This, however, is only one side of the story. Multinational companies have equally found that to enter the region, they require relationships with dominant domestic operators and access to these regional business networks. Thus, in the media industries there is a significant amount of jockeying taking place as various players try to work out who will be successful and why; who they should be teamed up with, and what they should be willing to commit. For some companies the decision will prove decisive.

The Overseas Chinese

The Chinese business networks in Asia rely heavily on family ties and personal relationships. Many of the companies which demonstrate the value of these relationships — Li Ka-shing's Hutchison in Hong Kong, the Salim Group in Indonesia and Shinawatra Computers & Communications (SCC) in Thailand — have recently moved into media ownership. However, while the access these relationships ensure should not be undervalued, neither should they be exaggerated. Although the networks exist, interfamily linkages are not as clear as has sometimes been suggested. (See Hamilton, 1992; Redding, 1990.)

The Charoen Pokphand (CP) Group, for example, is a relatively new entrant into media (and telecoms) businesses. The CP Group began as a major Thai-based agro-industrial firm but has diversified into a number of other areas such as petrochemicals and telecommunications and, more recently, into the broadcasting industries in Thailand. The group's various relationships with the Thai authorities helped to win it the TelecomAsia franchise (a joint-venture with the American telecommunications firm, Nynex) to build and operate two million digital lines in Bangkok, as well as the right to operate the first fibre optic cable television system in Thailand. The company also holds an interest in the Apstar1 and Apstar2 satellites, in association with Singapore Telecom, and several Chinese companies and ministries.

The CP Group has used this platform to expand regionally, with the biggest push to date being into China, including one of the first foreign investments in cable television developments in China (named Thai-China Cable Television) in the northeastern city of Shenyang, which was subsequently shut down by the Chinese.[31] The company also claims a controlling interest in Oriental TV, a television broadcaster in Shanghai. All of this at a time when the world's largest media combines are repeatedly refused entry into the China market. CP's agribusiness and related investments in China are very large, and it is speculated that the CP Group has become the largest foreign investor in the country.[32]

The CP Group began investing in China in Guangdong province in the 1980s. The core members of the group belong to the Chearavanont family, headed by the Jia brothers, who originally came from the Chiu Chow region of Guangdong (Suehiro, 1993). Family ties and regional associations are usually what is highlighted in the literature on the overseas Chinese business networks. While this *was* undoubtedly the case with the CP Group in the early days, they have since moved far beyond this, as the geographical diversity in the group's China operations now demonstrates.

Of greater asset in China is the ability of the overseas Chinese companies to operate in China's underdeveloped legal environment. A common language and culture helps. So too does the company's ability to strike up alliances with local officials. 'CP often cements with local officials by hiring them as either CP or joint-venture employees' (*Far Eastern Economic Review*, 21 October 1993, p.67). As a result of CP's China success, its TelecomAsia partner, Nynex, has been using the group's Hong Kong-listed subsidiary, Orient Telecom & Technology (OTT) to forge a quieter entrance into the Chinese telecommunications market.[33]

This again highlights the issue of localization, and a trend for local companies to expand into regional players. While it is likely that the dominant global companies — such as News Corporation and Time-Warner — will establish and increase their presence in Asia as a burgeoning middle-class begins to spend on entertainment and leisure services, it is also increasingly the case that, as with the telecommunications industry, a strong regional presence often requires a complex integration of global and local influences. It is within this regional vacuum that many Asian companies, such as TVB, are discovering a competitive advantage.

Television Broadcasts (Hong Kong) is controlled by Hong Kong's Sir Run Run Shaw and Malaysian-Chinese tycoon Robert Kuok (Kerry Group) who, in late 1993, also purchased a controlling interest in the *South China Morning Post*, the dominant English-language newspaper in Hong Kong.[34]

Like the CP Group, TVB has expanded regionally and internationally from a national base (Hong Kong) on the strength of strategic associations. At the global level, TVB is serving the overseas Chinese communities in North America, Europe and elsewhere. But it is in Asia (where trade in media products is dominated by Hong Kong programming) that TVB is strongest. Through its international arm, TVBI, it exports almost all of its locally produced period dramas, sitcoms, kung fu and police action series, an annual output of some 4,800 hours of Chinese-language programming dubbed into eight languages (Cambodian, English, Korean, Malay, Mandarin, Spanish, Thai, and Vietnamese) for distribution to more than 20 million people in 30 countries. Until recently, about one-quarter of TVB's revenue came from overseas distribution, comprising half its profits (*Asian Advertising & Marketing*, October 1991, p.12).

Malaysia, for example, is a particularly large market for TVBI in video distribution. TVBI runs, owns, and operates 200 video rental outlets in Malaysia. It is also expanding video operations in Singapore and Vietnam. To further this success and to tailor its products to individual markets, TVB shoots on location in countries such as Malaysia and Thailand. Other regional associations will result from TVB's plans to make co-productions with local TV stations, including Japan's NHK and several of the Chinese broadcasters.[35]

It is this well developed distribution network, the group's strong commercial contacts into China and the rest of the region, and its large Chinese-language programme library, which has made the company so attractive as a regional partner for the global groups seeking strategic alliances as a way into Southeast and East Asia. This was demonstrated in Taiwan where, to capitalize on the group's regional expertise, the international channels, HBO and ABN allowed themselves to be marketed as a TVBI package into the country's infant cable TV systems. In 1994, TVBI added more than 1,000 hours of Mandarin programming to its annual output for airing on TVBI's satellite service, the TVB 'Superstation' (TVBS), for redistribution into the Taiwanese cable television systems.

Case Study: Taiwan

Almost every home in Taiwan has access to television. The three commercial channels: Taiwan Television Enterprise (TTV), China Television Company (CTV) and China Television Service (CTS), are owned by the government, the ruling political party and military respectively. All accept advertising, programming is censored, and foreign imports face restrictions in airtime. Also, by government rule, 45 per cent of programming must be news, culture and education.

By 1994, Taiwan had probably the most competitive television market in Asia. One of the richest nations in the region with by far the largest cable penetration (somewhere between 45 to 60 per cent of television households), Taiwan is also the market which most international programmers keep at the forefront of their Asian expansion plans.[36] TVBS launched their regional Superstation in Taiwan (a joint-venture between TVBI and ERA International, a Taipei-based media company), while STAR TV initiated their first subscription service for Taiwan.

Taiwan has also amply demonstrated the difficulties that international programmers face in Asia and the need to understand individual market structures and work with dominant domestic business networks. When STAR TV introduced its STAR Movies subscription channel in April 1994, local operators boycotted the service. Their problems stemmed from a distribution agreement which soured relations with the powerful local United Communications Group. When the agreement with UCG broke down, a rival local organization, 52 Union, attempted to become STAR's exclusive agent. When STAR refused, and instead attempted to work on its own, 52 Union became vocal in the campaign against STAR.

Taiwan's current push for deregulation in the media industry arose partially as a result of the lifting of martial law in 1988. Restrictions on mass communications, including newspapers were loosened and the public began demanding the lifting of controls over TV stations. One result was the fantastic growth in illegal cable television services across the island, collectively known as the 'Channel 4' services, because they were an alternative to the three legal broadcast networks. Some 300 Channel 4 operators, broadcasting packages of approximately 20 channels (including the three free-to-air stations, illegally copied videos and down-linked satellite transmissions), would charge US$20 to 28 per month (see Table 6.5).

However, in early 1993 Taiwan's Parliament was forced, reluctantly, to approve a copyright agreement with the United States after trade officials were warned that Washington would impose general trade sanctions against Taiwan if the pact was rejected. Under the agreement, penalties were introduced to curb the growing export business in pirated software (*Business Times*, 20 January 1993, p.13).

Although US interests, represented by CNN and the Motion Picture Export Association of America (MPEAA), won their battle for copyright protection and a reduction in the local programming quota for cable, the final cable bill has also kept foreign incursions at bay by restricting foreign companies from holding interests in licensed cable operators. Under the 1993 Cable Television Law, companies such as the large American cable company, United International Holdings (UIH), which had been lobbying for foreigners to be able to directly participate in the construction of the cable television market in Taiwan, have been restricted to a 20 per cent share of a cable station and investment in only one station per district. Given that foreigners are allowed to own no more than 20 per cent of a Taiwan company which has holdings in cable stations, they are in fact limited to just under 4 per cent investment in any single cable operation.[37]

By the early-December 1993 deadline, some 618 renegade cable TV companies had registered with the Government Information Office. At that

> stage, the companies all received a one-year operating licence, while the government put the process into the next stage of review. The government then began cutting the wires of the companies that had not. Almost a year later the government received 204 formal licence applications.
>
> Taiwan's Cable Act provides for the establishment of 52 districts in Taiwan, each of which can have up to five licensed cable television providers, or in other words, some 260 potential operators. To the uninitiated, and according to the Parliamentary Yuan, this is a model for competition. Realistically, few groups will be large enough to bid for a licence in more than a handful of districts, but the few that can will become dominant through the economies of scale their networks allow. Furthermore, while there are 260 *potential* licences on offer, the government will award far fewer than that, since many districts across Taiwan will be neither large nor wealthy enough to justify more than two cable TV providers. Indeed, according to some industry representatives, there are some districts where the required investment will be a questionable proposition, in which case the job may be left to the government.
>
> Significantly, it is the government that is providing the distribution mechanism — the fibre optic backbone — across the island and *then* allowing licensed operators to use capacity to deliver programming within their franchise area. Given the opposition party's complaints as to media bias in political campaigning, the new system will leave the government in a remarkably strong position to continue to influence mass media communications on the island.[38] In 1992, the ruling party Kuomintang set up a company to buy up existing illegal cable companies.[39]

In South Korea the multinationals face a similar story — a result of the chaebol. The Korean government, faced by an invasion of foreign programme providers, realized that it would be difficult to turn back the clock and outlaw satellite dishes. So, as has been the case in Singapore, Malaysia and China, in an effort to wean South Koreans away from satellite television, the government turned to cable. Therefore to become successfully involved in pay TV developments the multinationals must create a successful partnership, in turn requiring an understanding of which domestic group is likely to be successful and why.

The chaebol are interested in the pay TV business largely because it would allow them to enter the very lucrative entertainment business. (Entry into the entertainment business is seen to complement existing interests in the electronics industry.) Korea's pay TV market has been estimated to be a 5 trillion won, market by the year 2000, when over 200 systems will be in operation. The hardware market will be worth an estimated 1.5 trillion won, software 40 billion won and advertising revenue and fees are expected to exceed 700 billion won a year (*Business Korea*, April 1992, p.8). As a result, most conglomerates have focused on programme distribution, which has become an area of fierce competition. The groups interested in becoming involved in cable TV service provision read like a 'Who's Who' of Korean business (Table 6.9).

Table 6.9
Korean Business Networks Involved in Cable TV

Group	Associates/Subsidiaries	Plans	Foreign associations
Samsung Group	Cheil Communication, involved in programme production and distribution	Looking for a way to match VCRs and other electronic hardware with software entertainment, much as Japan's Sony and Matsushita have done.	Samsung signed a contract with Walt Disney in 1990 to distribute Disney's movies and video programmes.
Lucky-Goldstar Group	The group's advertising firm, LG Advertising Co., has plans for programme production and distribution	Has operated a CATV system within its HQ buildings in Seoul since 1987 in oder to gain experience. Goldstar is concentrating on production and distribution; Goldstar Information and Communications is involved in system manufacturing.	The group is collaborating with General Instruments of the US.
Daewoo Group	already has distribution reputation in video market, in cooperation with Sejin Movie Co; is regarded as the forerunner in programme distribution due to relations with foreign programme suppliers.	Has started manufacturing cable-TV systems in a technical association with US-based company.	Has concluded agreements with RCA-Columbia, MGM and 20th Century Fox to purchase movies.
Hyundai Group	the group's cable capability is based through its subsidiaries, Hyundai Electronics and the Hyundai Theatre Co.	Has established a media business development department	Planning to boost its programming base by signing contracts with the BBC and ITN, among others.
Seoul Telecom Co.			Concluded a contract in 1990 to receive CNN and subsequently began a movie channel, TBS-TNT, a CNN-affiliated cable TV network; has also established links with ESPN.

Case Study: South Korea

Television penetration in South Korea is almost 100 per cent, but up until late 1991, Koreans had extremely limited choice with the existing four networks, which are state-owned and government-regulated. In 1991, Korea's first private TV channel, Seoul Broadcasting System (SBS), was started. The three state-run broadcasters are RBS-1, KBS-2 (Korea Broadcasting System) and MBC (Munha Broadcasting Corp.), and they cover the country through 83 affiliated stations. The fourth network is Educational TV which is non-commercial.

The Korean Cable Television Law (CATV Law) and Implementing Regulations (Presidential Decree 13682) went into effect on 1 July 1992. Initially, because of the government's concern about 'ethnic purity', no more than 30 per cent of programming on cable TV was to be foreign.[40] Also, foreigners were prohibited from participating in the cable TV broadcast business, programme supply and transmission. However, the foreign programming rules were changed with the passage of the draft cable bill so as to allow foreign companies to own up to 15 per cent of cable TV programme suppliers and lifting the level of foreign programming allowed from 30 to 50 per cent.

In a further bid to control developments, the law stipulated that conglomerates and media, including newspapers and broadcasters, were not to be allowed to operate pay TV, and that the maximum equity investment allowed by a single entity could not exceed 30 per cent of the value of a cable television company. This latter provision is included so as to restrict the chaebol.[41]

The government's cable television plan is to award 116 cable franchises covering the entire country, with the first 57, covering urban areas, awarded in early 1994. (Seoul alone will have 21. Other areas include Pusan, Taegu, Kwangju, Taejon and Inchon.) The network is to be rolled out by 1995 with cable-laying contracts going to Korea Electric Power Corporation, Korea Telecom and Data Communication Corp. (of Korea). As a first step in this process, programme supply contracts for the forthcoming cable TV networks were awarded by the South Korean Information Ministry to 20 companies in September 1993.

Korea Telecom has said that it will invest US$1.27 billion to establish its national cable TV system. But the company has also put the industry on notice that it wants to recover costs within 15 years. In addition to building the network, it will be responsible for transmitting programming, collecting subscriptions from the public and the redistribution of revenues to the cable system operators and programme providers.[42]

1. For an account of developments focussing more on the print and publishing side of the media in Asia, see Freedom Forum, 1993.
2. Chambers English Dictionary, Edinburgh, Scotland 1990.
3. For an interesting look at the convergence of linkages from the software or content side, see Doyle, 1992.
4. The difficulties in realizing business syneries on an international scale include restrictions imposed on cross-border mergers and acquisitions by governments wanting to protect their domestic industries.
5. The government later relented, requiring most advertising revenue to come from outside Hong Kong. For similar reasons, Wharf Cable — Hong Kong's cable TV operator — is similarly prohibited from carrying advertising until at least October 1996.
6. SCC owns 57 per cent of IBC, while the Mass Communication Organization of Thailand (MCOT) and private interests hold 43 per cent.
7. 'I keep hearing this word, like some sacred mantra: localise, localise, localise . . .' Bill Hooks, Managing Director, Home Box Office International, at the AIC Pan-Asian Satellite and Cable Television Conference, Hong Kong, 22-24 March 1994.
8. In 1993 CNN admitted to regret over a missed opportunity, wistfully reflecting on the fact that it was CNN that was first approached by STAR TV, 'We gave the BBC the opportunity to compete.' (*The Asian Wall Street Journal*, 22 March 1993).
9. The wording at the time was quite specifically, 'triumvirate', and not 'consortium', since 'consortium' could have provoked American anti-trust action.
10. The AsiaSat serviced area is divided between two 'footprints', a northern beam and a southern beam. The southern beam covers countries such as India, Pakistan, Indonesia and the Middle East. The northern beam straddles the predominantly Chinese speaking world of China, Taiwan, Hong Kong and Korea. In mid-1994, Rupert Murdoch suggested that one reason his company had dropped the BBC from the STAR TV service was to pacify the Chinese authorities who had not appreciated the BBC's coverage on human rights issues and its inclination to broadcast shows which the authorities found offensive, such as a controversial documentary on the late Chinese leader, Mao Zedong.
11. HutchVision's start-up budget for STAR TV was some US$300 million. Most of the advertisers who initially invested in the service were powerful overseas Chinese families with strong connections to the head of Hutchison, Li Ka-shing. They were invited by the Li family to invest, not as advertisers, but as part of a special investment package, the Hutchison Foundation Plan, also known as the '2 million club', for the amount each 'partner' invested.
12. 'The first group of channels that launched were free-to-air to create a demand or the product. The second group will build an infrastructure of subscription marketing and retrieval throughout the region of particular importance because

STAR's existing networks don't have a subscription revenue stream. The third step, when the AsiaSat-2 satellite is launched in 1995 will offer country-specific and more language-specific programming.' Gary Davey, STAR TV (*Multichannel News*, 3 August 1994)

[13] Neither the July 1992 figures nor the February 1993 figures, for example, included countries like Singapore and Malaysia (where STAR TV was not officially supposed to be received), Cyprus, Australia and Vietnam, from where STAR TV had been receiving faxes for program guides (*South China Morning Post*, 17 July, 1992). Moreover, the impressive November 1993 figures (42.7 million households) are for 13 countries, out of a possible 53, given this is the number of countries the C-band footprints encompass.

[14] STAR claims that the further localization of their music channel, now known as Channel V and with an increased Mandarin-language format, has led to a 108% increase in viewership.

[15] This position began to change in 1994 when the Japanese parliament for the first time allowed Japanese satellite systems to broadcast internationally.

[16] While Malaysia has begun to relax its attitude to satellite services, such that residents will be allowed to recieve Ku-band satellite signals when Binariang begins to broadcast from Measat, the government announced in late 1994 that they would be setting up a special unit to monitor all TV programmes aired by foreign stations. Monitoring is to be focussed on documentaries about the country and news bulletins, so as 'to prevent the integrity and image of the country and its government from being slurred.' *Asian Advertising & Marketing*, 18 November, 1994.

[17] Similarly, in Singapore, there has been the limited corporatization of Singapore Press Holdings and, in the near future, Singapore Broadcast Commission.

[18] Currently Binmantara holds 65 per cent of RCTI's shares, while the remaining stakes are held by individual holders including Suharto's son-in-law Indra Rukmana.

[19] Together, the consortium will pay the government an annual premium of US$8 million for the next ten years, beginning in 1994. TV4's broadcast area will be Klang Valley, comprising the predominantly middle-class regions of Kuala Lumpur and its neighbouring satellite city, Petaling Jaya.

[20] TV3 was launched in 1984, backed by the Fleet Group, the owners of the English-language newspaper, *New Straits Times*. TV3 is now moving towards national coverage: by 1993 it covered more than 90 per cent of the country.

[21] Binariang, a consortium controlled by Malaysian gaming and property tycoon Ananda Krishnan, intends to send its first satellite aloft in 1995, claiming to have incorporated high-powered 'video compression' features that will allow the service to carry a wider array of television channels. The fact that Malaysia is split into two land masses means that satellite capacity is used for real-time distribution of programmes for terrestrial re-broadcast. Currently, TV1 and TV3 use capacity on Palapa B2P for programme distribution from the Malaysian

peninsula to Sabah and Sarawak. These services will be expected to migrate across to the Binariang satellite once it is up and transmitting.

22 It is also the only country in this study to have adopted the NTSC 525 standard, as opposed to PAL 625.

23 It is interesting that, given developments in other media markets in East Asia, one of the prime considerations for increasing local production has been to 'stave off the threat of both cable and satellite television which cater to a relatively upscale market craving for foreign produced shows.' Eugenio Lopez, General Manager, ABS-CBN, speaking at the Philippine Cable Television Conference, Manila, 20–23 March 1994. A further factor noted by participants is that the instability of the peso has necessitated an increase in local programming.

24 These are companies that have in many cases been run by the People's Liberation Army (PLA).

25 For a useful historical overview of mass media developments and policy in China, see Chang, 1989.

26 ESPN later shifted to Hong Kong having been unable to conclude a deal for distribution with Singapore CableVision (SCV), and unhappiness over signal quality from the Malaysia Telekom facility. (*South China Morning Post*, 4 April 1993)

27 While this exemption — known as an item (k) exemption — has been useful, it is no longer valid since passage of the new Broadcasting Law, which became effective in early-1995. It remains unclear if the Chinese will permit this to be incorporated in the law.

28 It should also be noted that while the Discovery channel has decided to uplink from Singapore, it has established its regional office in Hong Kong.

29 In stark contrast to the rest of the region, Singapore has no import restrictions on commercials. This has helped it to develop an audiovisual industry that, by 1992, encompassed eleven film companies, 54 production houses and 25 sound recording studios. There is, however, an advertising ban on tobacco and censorship of sex, violence and language.

30 On 1 January 1994, the government launched a regional satellite television service, Singapore International Television (SITV), to provide an hour's worth of programming per day. Beamed to the region from Indonesia's Palapa satellite, the service has been set up by the Singapore International Foundation to promote the image of Singapore throughout the region. The service, to slowly expand in both time and intent, was broadcasting nine hours daily by February 1993, in English, Malay and Chinese, to a target audience living within a 1600 km radius of Singapore. Much of the proposed programming is to have a 'strong ASEAN flavour'.

31 This venture was subsequently closed down by Chinese authorities, reputedly because of the large size of the foreign involvement. As a result, the CP Group can be expected to be quieter about developments in the sensitive Chinese media industry.

32 Although as one report suggested, to value CP's China investments 'turns into an unproductive numbers game. Much of the family's business is channelled through private companies operating with the secrecy typical of many overseas Chinese conglomerates.' (*Far Eastern Economic Review*, 21 October 1993)

33 In 1993, Nynex purchased a 23.1 percent stake in OTT, 'on condition OTT undertook to reduce its non-telecoms assets "as and when appropriate"', according to a report in the *South China Morning Post* (18 September, 1994). The report continues, 'Nynex [wants] OTT to concentrate on telecommunications business in China. . . . the share deal also stipulated OTT would make no purchases outside the telecommunications field without the prior approval of Nynex.'

34 In late 1993, Rupert Murdoch's News Corporation sold its controlling interest in the *South China Morning Post* so as to concentrate on satellite broadcasting via the STAR TV entity, of which the group had purchased 63.5 per cent for US$525 million in July 1993. In an interview, Murdoch later declared that he had sold the newspaper share to avoid the possibility of having journalists antagonize authorities in China, and then having this rebound against the satellite broadcaster.

35 As of late 1994, TVB had programme licensing agreements with at least ten mainland Chinese television stations.

36 Industry officials have announced at regional conferences that they expect approximately 5 million households to eventually subscribe to one cable system or another. With a monthly subscription of between US$22 to 24, that translates into total revenues of over $100 million a month.

37 The law does, however, create significant business opportunities for US cable TV equipment suppliers, programme providers and network design and engineering companies.

38 Members of the opposition Democratic Progressive Party (DPP) claimed that one significant reason that they failed to win enough seats to challenge the government outright in the 27 November 1993 local elections, was the biased reporting on the three government-controlled broadcast station. The DPP has attempted to fund and establish an illegal alternative, the National United Television (NUTV) network.

39 The only group of comparable size is the Koo family, reputedly Taiwan's second richest. C.F. Koo is a long-standing member of the ruling party's Central Standing Committee.

40 In some categories the figure was set even lower: no more than 25 per cent of a music channel's programming should be foreign. The South Koreans, like the Taiwanese, have waged a battle against Japan's NHK satellite television services being shown in the country on the grounds that it poses a threat to the indigenous culture. That so many South Korean viewers tune into Japan's NHK — in a country where Japanese films, video, theatre and magazines are outlawed — appears to have fuelled the government's determination to do

something about what it sees as an invasion of the airwaves by foreign programmers.

41 The chaebol are also specifically prohibited from operating news channels.

42 South Korea hopes to launch its first satellite in 1995, with a second one in late 1995. Korea Telecom is the major investor in the US$1 billion Koreasat project. In addition to capacity for 5,300 voice circuits, there are facilities for three television broadcasting channels and four video relay channels. When operational, Koreasat is to be used both for DBS broadcasts and to deliver programming to the cable networks.

Asia – The Supplier's Dilemma

Nick Ingelbrecht

Asia: Success and Failure in Diverse Markets

For the world's leading telecommunication equipment suppliers, Asia has become of critical importance to their future growth and profitability. The number of telephone main lines in Asia[1] grew by 11 per cent in 1993 — the fastest of any region in the world — while mobile communications equipment revenues grew by 64 per cent the same year. Capital expenditure on telecommunications in the Asia-Pacific region as a whole (including Japan) totalled US$57.4 billion in 1993, a 2.9 per cent increase on the previous year despite falling equipment prices, according to the International Telecommunication Union. (See Table 7.1 which indicates that the Asia Pacific Region as a whole accounted for 77 per cent of the 31 million main telephone line growth 1992–93.)

> The Asia Pacific region continues to be the fastest growing telecommunication region, with China contributing almost a fifth of the world's new main lines in 1993. (ITU/94–18, *Global Telecommunications Shows Mixed Growth in 1993*).

As a measure of its market potential, the Asia-Pacific region accounted for just 23 per cent of the world's telephone lines in 1993, despite being home to 58 per cent of the global population. Against this, spending on telecommunications equipment in the Asia-Pacific is predicted to outstrip that of any other region until the end of the decade, annually around US$20.2 billion at 1993 prices. However, this will only increase teledensity — the basic indicator of main telephone lines per 100 population — to 6.2 per cent,

Table 7.1
Asia-Pacific: Main lines, Telecom Revenues and Investment in Telecoms

	1992 Total	1992 Share	1993 Total	1993 Share
Main lines (million)				
World	576		607	
Asia-Pacific	125	22%	138	23%
— Less Japan	68	12%	80	13%
Main lines added 1992–1993 (million)				
World			31	
Asia-Pacific			12	40%
— Less Japan			12	37%
Telecom revenues (US$ billion, current prices)				
World	445		445	
Asia-Pacific	94.1	21%	108.3	24%
— Less Japan	39.5	9%	42.4	9%
Telecom capital expenditure (US$ billion, current prices)				
World	130		125	
Asia-Pacific	38.7	28%	37.4	30%
— Less Japan	17.1	13%	20	16%

Source: ITU.

compared with the 40 to 50 per cent penetration levels in developed countries. Such a level of investment nonetheless represents more than one-third of the total world spending on telecommunications. If the projected spending for Oceania is included, then the Asia-Pacific region as a whole will account for nearly one half of world telecommunications spending by the end of the decade (Table 7.2). The growth of Asia's mobile equipment market has been even stronger. (Figure 7.1 shows a worldwide increase of 11

Table 7.2
Projected Investment and Growth in Telecommunications

	Main telephone lines per 100 inhabitants 1992	Main telephone lines per 100 inhabitants 2000	Investment required (1993–2000) Total US$m	Investment required (1993–2000) Per year US%m	main lines to be added (1993–2000) Total 000s	main lines to be added (1993–2000) CAGR %
Asia	3.86	6.19	161,327	20,165.9	107,548	7.1
Africa	1.47	2.56	18,414	2,301.8	12,282	9
Americas	26.04	31.13	105,117	13,139.6	70,083	3.7
Europe	35.35	47.25	134,050	16,756.3	89,364	4.8
Oceania	36.79	45.03	5,911	738.9	3936	4.2
Former USSR	13.84	21.79	41,335	5,166.9	27,559	6.7
World total	10.48	13.87	466,154	58,269.3	310,772	5

Source: ITU.

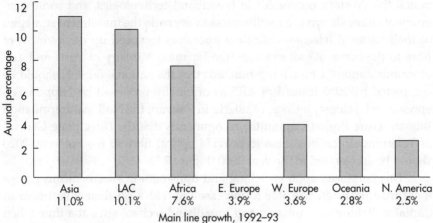

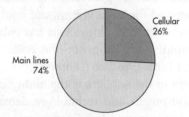

New telephone access points 1993: 42 million added

Note: LAC = Latin America and Caribbean. E. Europe includes Central Europe and former Soviet Union.
Source: ITU.

Figure 7.1 Telephone main line and cellular growth, 1993

million cellular telephone subscribers.) Mobile equipment sales topped US$7 billion in 1993, an increase of 64 per cent on the previous year.

Asia currently boasts five of the world's top ten fastest growing telecommunications markets in terms of fixed line growth: China, India, Indonesia, Thailand and Malaysia. Asia also boasts some of the most sophisticated and mature telecommunications markets, such as Japan, Hong Kong, Australia and Singapore. Asia is also a highly concentrated market. By 2010, 90 per cent of the region's expenditure on equipment and services will be attributed to six countries: Japan, China, Australia, Taiwan, the South Korea and India. (See BTCE, 1991.) Increasingly countries have come to recognize the imperative of telecommunications for sustained economic growth, and most now acknowledge the need for rapid telecommunications infrastructure development if they are to avoid being left behind developed nations.[2] Asia's telecommunications infrastructure boom is not confined to the developing nations.

The mature economies of the region are well aware of the need to match the Western economies in broadband technologies, and some governments have shown their willingness to overrule the investment strategies of their national telecommunication operators to speed up deployment of fibre to the home. As an example, the Japanese Ministry of Posts and Telecommunications' Council recommends that the national B-ISDN should be completed by 2010 instead of 2015 as originally proposed by Nippon Telephone and Telegraph Corp. Similarly in Taiwan, the National Information Infrastructure Project Committee recommends that the Directorate General of Telecommunications' plan to provide optical fibre to the home by 2020 should be accelerated to the year 2000 (Figure 7.2).

This emphasis upon switching and transmission speeds means equipment manufacturers are able to increase the return on their investment in digital switching transmission and broadband technologies at a time when they are being obliged to invest ever larger amounts in research and development.[3] Hence the Asian market provides suppliers with large growth potential, immediate demand for large volumes of equipment and a growing appetite for state-of-the art technology. This has enhanced the ability of suppliers to support rapidly declining per-line costs, particularly in the region's biggest market, China. (See also Chapter 2.)[4]

Of increasing concern to the supliers is the ability of developing countries to pay for their equipment and technology demands. This issue has become more pressing because supplier finance and government-assisted preferential loans have long provided an important key to market entry in developing countries.[5] Customers are also obliged to match their technology choices with the available sources of finance and concessionary funding often constrains a national operator's choice to a supplier from the country providing the loans. Developing countries may therefore find themselves in a position of chasing the finance rather than the technology most suitable to their needs, resulting in a proliferation of different types of equipment from different suppliers and fostering a short-term outlook in the relationship between suppliers and their customers. Technology choices are often far from straightforward, particularly where trade-offs between functionality and cost have to be weighed. In the provision of basic infrastructure, the latest technology may be the most cost effective in absolute terms, but not if vital components and spares need to be imported and paid for with scarce foreign currency. Intermediate technology may provide a better solution for developing countries if engineers are cheap and plentiful, and civil works subsidized by other government departments.

Indeed, the merit of 'technology leapfrog' by developing countries is frequently promoted as an objective in itself, but the consequences are often ignored. An order by the former Marcos government in the Philippines for the national carrier to install only digital switching equipment in favour of

Asia — The Supplier's Dilemma   197

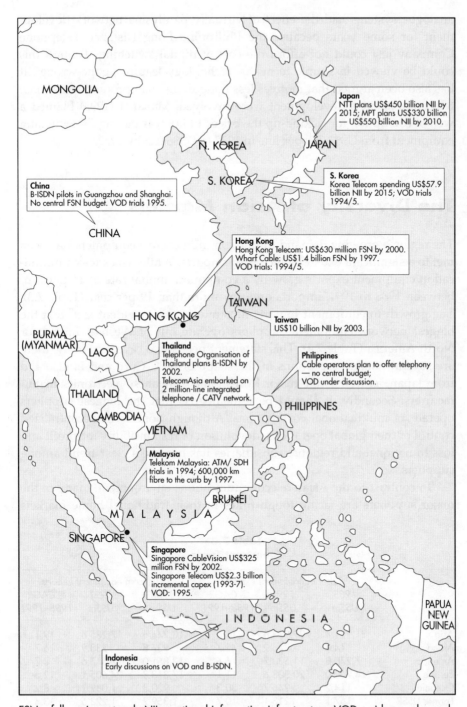

FSN = full service network; NII = national information infrastructure; VOD = video on demand.

Figure 7.2 Technology convergence in Asia — national information infrastructure projects around the region

analogue systems had the effect of virtually paralysing network development for some years because the Philippine Long Distance Telephone Company just could not afford the cost of digital switching. Though this could be viewed in simple terms as 'technology leapfrog gone-wrong', it has also been argued that motives less honourable than promoting national telecomunications development were involved. Manapat (1993) blamed a web of corruption for jacking-up the prices PLDT was paying for switching equipment from US$1,750 per line in 1977 to US$6,023 by 1982.[6]

The Diversity of Asian Markets

The region's growing demand for telecommunications equipment has been met to a significant extent by increased imports. While Asia's telecommunication equipment exports grew by a compound annual rate of 11 per cent between 1988 to 1991, imports grew by more than 19 per cent (Table 7.3). The growth in equipment imports illustrates a fairly evident fact, that the biggest network equipment suppliers operate out of Western Europe or North America (Table 7.4). This statement needs two caveats: Firstly, there are several notable exceptions in the shape of NEC Corp and Fujitsu Ltd from Japan, as well as the major Korean and Taiwanese electronics manufacturers. Secondly, it should be remembered that the biggest suppliers operate as multinational corporations. Although none have decentralized control of their global operations, devolution of responsibility for profit and loss to national and regional subsidiaries has become a clear trend among suppliers.

In contrast to the Asian telecommunications environment, many of the major suppliers are facing tough times in their traditional home markets

Table 7.3
Equipment Trade

	Telecom equipment exports			Telecom equipment imports		
	1988 US$m	1991 US$m	% CAGR 1988–1991	1988 US$m	1991 US$m	% CAGR 1988–1991
Asia	21,921.4	29,973.8	11	10,244.6	17,235.6	19.1
Africa	14.4	46.2	42.7	937.8	1238.9	16.7
Americas	7,270.8	11,533.7	16.6	13,851.0	17,617.6	8.7
Europe	14,960.0	20,508.6	11.1	17,313.8	24,605.1	12.5
Oceania	114.2	246.7	30.3	920.3	1,089.1	8.2
Former USSR	157.1	213.1	10.7	1,359.2	1,223.5	-3.4
World total	44,438.0	62,522.1	12.1	44,626.8	63,009.9	12.5

Source: ITU.

and this has inevitably influenced their global strategies. All of them have been grappling with spiralling research and development costs as well as tougher competition and lower profit margins. While the mature telecommunications markets of the USA, Europe and Japan have suffered a slow-down due to the effects of economic recession, the reverse has been true in Asia. Based on information supplied by public network carriers, in a press release the ITU (ITU94–18) reported that the world market for telecommunications equipment did not grow at all between 1992 and 1993, with revenues remaining stable in 1993 at US$120 billion. For their part, the majority of the big switch vendors blamed flat revenue growth on increased competition in international markets, while according to *Communications Week International*, 14 November 1994 (see Table 7.4), suppliers of data communications equipment and software, such as Cisco Systems and Cabletron, reported the fastest growth rates. The major vendors have consequently turned their attention outwards from their traditional home markets, and are intent on addressing new markets, both geographical and technological.

In developed countries, the users themselves are becoming an increasingly important source of new revenue,[7] while liberalization and the convergence of communications technologies is creating new equipment markets in developing countries, such as the mobile communications business and a regional cable TV industry. (See Table 7.5 and compare Table 6.2 for growth rates.) Prior to India's recent moves to liberalize the telecommunications market, for instance, mobile phone services were not available at all in the country. Cellular phones were viewed as expensive luxuries which India's hard-pressed foreign currency reserves could ill afford.

In simple commercial terms, Asian markets are often portrayed as green field sites of opportunity for suppliers. In practical reality, many of the region's brightest telecommunications prospects are walled-in with market access restrictions and infested with quagmires of political intrigue and vested financial interests. In addition, Asian countries sometimes lack the basic skills and experience in technical and commercial planning necessary to develop cohesive investment programmes:

> Further, they lack the necessary knowledge and experience to plan and specify the equipments needed to meet the objectives of any well-formulated plan. In an environment of a wide range of network and technology options available, the differences between well-planned and specified telecommunications services and inadequately planned systems are reflected in significant large scale differences in investment capital and commercial returns for the sector.
> (ITU, 1993, *Pipeline Project RAS/92/PO4, Guidelines for the Preparation of Investment Programmes and for Procurement*)

Table 7.4
Top 50 Communications Equipment Manufacturers

Rank 93	92	Company	Communications equipment revenue ($m 1993)[1]	Change in communications equipment revenue (1992–93)[2]	Total revenue ($m 1993)	Change in total revenue (1992–93)
1	1	Alcatel Alsthom (France)	14,544	−2.1%	27,605	−3.3%
2	2	Siemens (Germany)	11,986	+6.8%	49,385	+4.0%
3	3	AT&T (U.S.)	11,783	+9.0%	67,156	+3.5%
4	4	Motorola (U.S.)	10,105	+21%	16,963	+28%
5	7	NEC (Japan)	8,714	+0.8%	32,192	+1.8%
6	5	Northern Telecom (Canada)	7,861	−21%	8,148	−3.1%
7	6	Ericsson (Sweden)	7,703	+34%	8,088	+34%
8	8	IBM (U.S.)	5,300	0%	62,716	−2.8%
9	9	Fujitsu (Japan)	4,388	+3.1%	28,231	−9.3%
10	10	Bosch Group (Germany)	2,655	+4.4%	19,655	−5.6%
11	17	Nokia (Finland)	2,161	+51%	4,418	+30%
12	18	Matsushita Electric (Japan)	2,046	+1.1%	59,565	−6.1%
13	11	GEC (U.K.)[3]	1,917	−15%	14,571	+3.1%
14	13	Philips (Netherlands)	1,831	−12%	31,672	+0.5%
15	26	Samsung Electronics (South Korea)	1,788	+52%[4]	10,159	+34%
16	20	Toshiba (Japan)	1,572	−8.0%	29,027	−0.6%
17	12	Italtel (Italy)	1,558	−11%	1,670	−11%
18	16	Ascom Group (Switzerland)	1,538	−14%	2,140	−6.2%
19	19	Matra Hachette (France)	1,508	−0.1%	9,532	−2.0%
20	23	Oki Electric (Japan)	1,462	+1.2%	5,859	+1.8%
21	22	Sumitomo Electric (Japan)	1,453	−3.5%	9,906	−3.1%
22	15	GM Hughes (U.S.)	1,430	−25%	13,518	+9.9%
23	21	Hitachi (Japan)	1,429	−6.0%	66,549	−1.8%
24	24	Mitsubishi (Japan)	1,377	−2.0%	27,927	−4.8%
25	28	Japan Radio (Japan)	1,316	+14%	2,002	+1.5%
26	42	Cisco Systems (U.S.)	1,243	+92%	1,243	+92%
27	–	Martin Marietta (U.S.)[5]	1,135	NR	9,435	+58%
28	35	General Instrument (U.S.)	1,125	+33%	1,393	+30%
29	32	Novell (U.S.)	1,123	+20%	1,123	+20%
30	33	Sony (Japan)	1,089	+4.0%	33,577	−6.5%
31	25	Sagem (France)	1,049	+2.5%	2,302	+6.4%
32	30	Corning (U.S.)	1,005	+2.1%	4,005	+7.9%
33	31	Harris (U.S.)	906	−5.0%	3,099	+3.2%
34	45	3Com (U.S.)	827	+34%	827	+34%
35	35	BICC (U.K.)	826	0%	5,885	+7.4%
36	44	Hewlett-Packard (U.S.)	780	+24%	20,317	+23%
37	37	Pirelli (Italy)	768	+3.0%	5,876	+12%
38	41	Sharp (Japan)	749	+0.4%	13,400	+0.8%
39	49	DSC Communications (U.S.)	731	42%	731	+36%
40	40	Ricoh (Japan)	731	−7.2%	8,708	−5.2%
41	38	DeTeWe (Germany)	721	−5.2%	864	−1.2%
42	–	SynOptics (U.S.)	704	+81%	704	+81%
43	48	Sanyo Electric (Japan)	701	+10%	13,730	−0.7%
44	46	Uniden (Japan)	699	+4.8%	699	+4.8%
45	39	Racal Electronics (U.K.)	667	+0.9%	1,332	−2.4%
46	36	Furukawa Electric (Japan)	665	−30%[6]	5,157	−5.6%
47	–	Anritsu (Japan)	604	+3.0%	734	−0.5%
48	–	Cabletron (U.S.)	598	+43%	598	+43%
49	–	Fujikura (Japan)	595	−12%	2,247	−0.8%
50	–	Scientific-Atlanta (U.S.)	576	+33%	731	+26%

Notes: Figures are for 1993 fiscal year. Figures for Japanese and U.K. companies are for year ended 31 March 1994; Cabletron year ended 28 February 1994; 3Com year ended 31 May 1994; Cisco ended 25 July 1994. Harris figures are for year ended 30 June 1993. All figures converted to U.S. dollars at 1993 average exchange rates. [1]Communications equipment revenue estimates by Sirius. [2]Year-on-year changes calculated based on home

Asia — The Supplier's Dilemma   201

Table 7.4
(continued)

Income before taxes ($m 1993)	Change in come (1992–93)	Employees (1993)	Total revenue per employee	Main communications products
1,582	−7.4%	196,500	140,485	Public, private network systems
1,761	−8.9%	391,000	126,304	Public, private network systems
6,204	+4.1%	308,700	217,545	Public, private network systems
1,525	+91%	120,000	141,358	Mobile, data communications products
226	NR	147,910	217,648	Public, private network systems
(1,070)	NR	60,293	135,140	Public, private network systems
399	+150%	69,597	116,215	Public, mobile network systems
8,797	NR	256,207	244,786	Computer networking products & systems
236	NR	163,990	172,153	Public, private network systems
433	−42%	156,615	125,501	Public, private network systems
1,146	NR	25,800	160,785	Mobile, public network systems
1,153	−21%	254,059	234,452	Private network equipment
1,301	+0.1%	86,121	169,191	Public, private network systems
822	NR	238,500	132,798	Mobile, public network systems
262	+122%	44,733	227,119	Public, private network systems
811	+4.9%	175,000	11,705	Private network systems
37	−78%	16,254	102,897	Public network systems
(221)	NR	14,955	143,093	Public, data network systems
413	NR	41,904	227,468	Public, private, mobile network systems
12	NR	22,585	259,416	Public, private mobile network equipment
498	+4.4%	39,355	251,706	Cables
1,494	NR	78,000	173,308	Satellite, mobile products & systems
2,054	−2.7%	330,673	201,274	Public, private network systems
643	−10%	111,053	251,470	Mobile, private network equipment
50	−3.7%	3,933	325,353	Wireless networking equipment
145	+69%	2,262	549,503	Data networking systems
725	594%	92,000	102,554	Satellites
114	NR	10,100	137,871	Cable products
104	−72%	4,429	253,533	Computing networking software
919	+10%	103,000	258,282	Private network equipment
174	+8.5%	14,495	158,831	Public, private network systems
157	−54%	39,200	102,168	Cables
170	+36%	28,300	109,505	Radio equipment, components
20	−68%	1,971	419,581	Data networking systems
156	+65%	44,000	133,746	Cables
1,783	+35%	96,200	211,195	Data networking, test equipment
(32)	NR	53,540	109,749	Cables
549	+5.0%	42,883	312,484	Private network equipment
109	+548%	4,041	180,832	Public network systems
230	+52%	48,000	181,414	Fax machines
35	−28%	5,994	144,199	Public, private network systems
130	+94%	1,736	405,814	Data networking systems
29	+481%	59,624	230,280	Fax machines, phones
NA	NA	NA	NA	Mobile, cordless phones
40	−45%	11,500	115,846	Private network systems
101	−4.2%	NA	NA	Cables
21	+1.2%	NA	NA	Public phones, measuring equipment
183	+41%	3,500	170,889	Data networking systems
5	−90%	4,286	524,153	Cables
33	+52%	3,600	202,953	Satellite, cable products & systems

currencies. [3]GPT revenue included in GEC figures, not Siemens. [4]Jump in communications revenue partly due to revised estimates. [5]GE satellite business sold to Martin Marietta. [6]Drop in communications revenue partly due to revised estimates. NR = not relevant. NA = not available.
Sources: *Communications Week International*, 14 November 1994, p.22; Sirius, Montpellier, France.

Table 7.5
Asia's Cable TV Markets (by Year End 1994)

	Cable TV Households	Television Receive Only Satellite Dishes
China	32,000,000	3,000,000
Hong Kong	150,000	–
India	10,000,000	2,000,000
Indonesia	–	750,000
Japan	9,230,000	5,580,000
Malaysia*	–	40,000
Philippines	250,000	30,000
South Korea	6,000,000	500,000
Taiwan*	3,500,000	230,000
Thailand	126,000	–

* Taiwan TVROs notional, Malaysia TVROs illegal.
Note: Cable TV includes MMDS (Microwave Multipoint Distribution System)
Source: Cornerstone Ltd, MPT Japan, ITU, independent estimates

Asia is clearly a fragmented market, with many regional differences in language, culture, ideology, income levels and income distribution, and the question of difference certainly extends to industrial issues such as type approval procedures, taxes and controls on imports and currency repatriation. Hence detailed generalizations about the telecommunications equipment sector in Asia are prone to contradiction and supplier strategies vary in different markets. Moreover, suppliers which have achieved notable successes in Asian markets appear to have done so because they left their accountants at home, and gambled on their ability to adapt to local market conditions.

Suppliers and the State

Northern Telecom, for example, claims its success in breaking into the Japanese public switching market resulted from its willingness to re-engineer its telephone exchanges to an extent that other suppliers were not prepared to do at the time. The company says that while other suppliers argued with NTT over the onerous nature of its product specifications, Northern Telecom adapted its product line to the Japanese requirements and as a result, was able to leverage a small switch sale into a US$270 million business within three years.[8] As another example, ITT Corp's willingness to transfer its digital switch production processes to China in the early 1980s and a subsequent decision to re-invest in a near-bankrupt production facility in 1986, earned what is now the Alcatel group a 49 per cent market share of the Chinese public switch market by the end of 1992 (Figure 7.3). To its lasting chagrin, AT&T passed up an opportunity to transfer its switching technol-

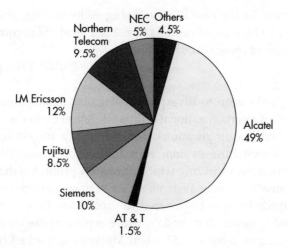

Figure 7.3 China's public switch market (total 43 million lines), 1992.

ogy to China in the early 1980s. The US company similarly decided not to enter China's radiopager market at an early stage. AT&T executives admit that in the light of hindsight, they may have acted differently.[9]

AT&T has faced a similar dilemma more recently, over the transfer of microelectronics production facilities to China. In order to secure a strong presence in China's telecommunications equipment market, the US manufacturer was offering to export its technology for producing sub-micron integrated circuits. But while this would help secure access to a strategically important market, it could affect customer demand for integrated circuits manufactured at other AT&T plants around the world. What these examples illustrate in practical terms, is that suppliers in increasingly competitive international markets are having to exact the highest possible price for transferring their manufacturing skills and research and development technology offshore, while retaining maximum control over the sources of production of vital components.

Asian governments have adopted varying approaches to the development of their telecommunications industries and their national aspirations do not always coincide with the commercial objectives of overseas suppliers. Controls on the flow of hard currency, local job creation and technology transfer are the bullet which overseas suppliers have had to bite in order to achieve access to Asian markets. In return Asian governments have had to reconcile national sovereignty issues with the need for foreign resources:

> Should a country share its domestic market with a foreign organization? Normally no entity would like to do this unless a portion of the market can be traded with other

tangibles. In the case of developing countries, the sharing would be because of capital to raise the level of telecommunications in those countries.

(Subhi, 1993, p.15)

A country's decision to invest in a particular manufacturer's network switch or to adopt a certain national cellular standard carries with it immense political and financial implications. For example, a project to introduce a second digital switching system into Indonesia prompted a series of byzantine political manoeuvring which ultimately prompted the government to adopt two new switching systems rather than one as originally tendered for. This compromise was engineered to protect the flow of economic aid from Japan and to ensure Indonesia's future exports to the US market.[10]

Similar concerns were voiced when Malaysia embarked on its current telecommunications expansion programme in 1992. The Ministry of Finance ignored the recommendation of Telekom Malaysia's own tender review board in order to accept bids from five different switch suppliers rather than the three favoured by the company. The Finance Ministry, which holds a 76 per cent stake in Telekom Malaysia, decided that a M$2 billion contract for four million exchange lines over five years should be divided between joint ventures representing Alcatel, Ericsson, NEC, Nokia and Fujitsu.[11] One noteworthy aspect of the denouement of this particular tender was the claim that the Finance Ministry rejected bids from Siemens and AT&T because of overzealous lobbying by politicians and businessmen. And while the Finance Ministry was also concerned to increase competition between suppliers and to reduce prices, Telekom Malaysia was worried that the proliferation of switch suppliers would generate a much higher cost overhead in terms of duplicate equipment inventories, training and documentation. Decisions of this nature inevitably distort the framework for future telecommunications projects since suppliers adapt their market entry strategies to take account of local conditions.

Whether such distortions in seemingly straightforward decision making processes are created by influence-pedalling political bickering, bureaucratic inertia or bribery, the result inevitably compromises the efficient build-out of the national telecommunications infrastructure. In the case of Indonesia, the whole process of acquiring a much needed alternative switching technology and the finance to go with it took more than two years to resolve and two rounds of tendering before the final contracts were awarded and the work itself was allowed to commence. In Malaysia, despite ministerial protestations to the contrary, subscribers will ultimately have to bear the additional overhead cost of a proliferation of digital switching systems as well as suffering the technical disadvantages involved in interworking between different vendors' exchanges.

What may appear as capricious decision making by Asian governments has done little to foster a spirit of generosity among telecommunication equipment suppliers. More often than not, commercial realities force suppliers to adhere strictly to the letter of their obligations, but little else. In Taiwan, for example, three public switch suppliers installed equipment for Integrated Services Digital Network (ISDN) pilot trials in 1991. ISDN-capable switches were installed in various parts of the island by the established joint venture suppliers AT&T, Alcatel and Siemens. Though the switches functioned well in isolation, ISDN interworking between the different manufacturers' exchanges was not possible. One switch vendor explained how suppliers had no commercial interest in modifying switch interfaces — the systems and hardware which enable a multiplicity of exchanges in a network to be connected together — nor the signalling software in order to accommodate other operator's equipment. The view expressed among suppliers was that:

> It is a question of us understanding their software and interface and them understanding ours. No one switch vendor wants to expend time and money on making modifications to other people's switches.
> (Interview, December 1993, with Taiwan equipment supplier, who requested anonymity.)

The Directorate General of Telecommunications (DGT) was required to hold a special tender to facilitate interworking between ISDN switches, which was accomplished by Siemens AG, prior to the launch of commercial services in 1994.

While the DGT's difficulties with ISDN relate to just three different types of digital switch in Taiwan, such problems are compounded in larger markets like mainland China where, at the last count, ten different types of digital central office switch were connected to the public network.

Recognizing this danger, China's State Planning Commission issued a regulation in 1989, titled Directive 56, which limited to three the number of foreign joint venture digital switch manufacturers permitted to set up factories in China. These comprised the Alcatel subsidiary, Shanghai Bell Telephone Equipment Manufacturing Co., NEC's joint venture switch company, Tianjin NEC Electronics and Communications Co., and Siemens' Beijing International Switching System Corp., also known as BISC. While Directive 56 had the commendable aim of trying to contain the interoperability problems posed by different manufacturers' exchanges, it also threatened to severely curtail China's telecommunications expansion plans. Put simply, China had manufacturing capacity for just over one million lines per annum in 1992, but was harbouring aspirations to install ten times this number. The State Planning Commission modified Directive 56 early in 1993 to permit AT&T and Northern Telecom to set up switch manufacturing operations in China. LM Ericsson has attempted to bolster its position as a switch supplier to China by setting up a production line for its AXE mobile switch at its Nanjing plant and Fujitsu is negotiating to do the same.[12]

With five joint venture public switch manufacturers now active in China, it is crucial for each of the major suppliers to secure sufficient market share to ensure future survival in the market. At the very least, once a province has adopted a particular system it is unlikely to dispose of its investment at an early stage in the interests of national uniformity. On a more positive note, provinces will also find it cheaper to upgrade equipment from their existing suppliers than to buy in a different vendor's technology. In this context, it is interesting to note Alcatel's development of an upgrade path for its time division multiplex digital exchange to the fast cell switching technique of asynchronous transfer mode. The company claims that for an incremental cost of 10 per cent of the price of the original TDM system, it will be possible to upgrade the kernel of the digital exchange to ATM working, with the switching software portable between either device. In addition, Alcatel has argued that by 1995, ATM would be able to justify itself as a cost effective means of switching basic telephony services rather than solely providing a Rolls Royce solution for delivering high bandwidth applications such as video and bulk file transfer.[13] If these developments prove to be commercially practical, their significance is that they will allow suppliers like Alcatel an added advantage in retaining market share in face of the major technology leap portended by (Broadband) B-ISDN.

Market Liberalization

On the surface, it appears that entrenched suppliers are in an extremely strong position against new competitors through their existing market presence. However, newcomers do have advantages of their own. They are able to offer the very latest telecommunications technology, while existing suppliers are saddled with maintaining a return on investment on the technology which they may have transferred three or five years ago. Newcomers may also bring with them fresh sources of government or supplier finance. This is an added incentive for governments to put the liberalization of equipment supply at the forefront of their telecommunications sector reform, because in doing so, they can open the gates to an influx of overseas finance and technology.

Market liberalization of equipment supply, particularly in Asia, has had a marked impact on manufacturers. Suppliers who previously operated hand in hand with government agencies or local telecommunication monopolies have found their once dominant market position being sliced up into smaller shares. Such a loss of market share generally has not proved disastrous for incumbent suppliers. Experience generally indicates that liberalization usually generates an expansion in the size of the market which will more than offset any loss of business.[14] In Europe and progressively in Asia, there has been a general breaking down of the traditional cosy relationship between incumbent suppliers and national phone monopolies. Siemens, for example, has been dislodged as the single supplier of digital switching equipment in Indonesia and the Philippines.

Asian telecommunication carriers are shopping around and sourcing equipment competitively from different suppliers — and manufacturers are increasingly able to sell equipment to new carriers in markets where competition has been introduced. The seven major switch suppliers have blamed their 1993/94 slow-down in profit growth on the effects of increased competition. Asian companies could follow the example of Alcatel SEL, the German subsidiary, which was losing over half a million US dollars a day in 1994 because its main customers at Deutsche Telekom and Deutsche Bahn were using other suppliers. Some of the Japanese electronics companies which have traditionally served NTT could also be vulnerable in this respect. Losing a cosy relationship in a vendor's home market has led many to seek out new markets abroad, a rationale which has driven many companies to globalize their operations.

Liberalization of the market for network equipment has come about only recently in Asia, but has been rapid and widespread. Indonesia, for example, has expanded its switch suppliers from one to three, Singapore is buying switching systems from Alcatel after being dependent on Fujitsu for a decade, and Thailand and Malaysia have each widened their markets from two to five switch suppliers. Some telecommunications specialists argue that this multiplication of suppliers has created a level of strategic uncertainty among manufacturers.[15]

Although localizing manufacture has become an inevitable part of most large switching contracts, so too has the desire of local partners and their governments to use their new facilities to generate exports. Against this, multinational suppliers have had to develop strategies to make their ventures in each country economically self-supporting. With the devolution of profit and loss accountability to each country, subsidiaries of the same manufacturer run the risk of competing for the same export markets with the same products. Since customers can obtain the same products from different sources, the inevitable effect is to reduce prices. A key issue for both switching suppliers and their customers in negotiating manufacturing agreements is the amount of local content and the level of technology transfer — from the manufacture of the microprocessors at one end of the scale and assembly of knock-down kits at the other:

> In the public switching industry, it is almost impossible to develop a new large switching system and compete in the world market. Muskens (1988) predicted that each manufacturer should have at least 16 per cent of world share in order to exist, because of high research and development cost. In other words, if the world market would be a single market, there might only be six manufacturers left. Therefore another approach to transfer of technology should be worked out. One possible answer is to focus the transfer only in the software area.
>
> (Santosa, 1993, p.1055)

With Asian telecommunications markets expanding at such a rapid pace, the major equipment suppliers have recognized the strategic necessity of building up market share at an early stage in order to lock customers into their particular systems. On the one hand, suppliers are often prepared to offer very attractive up-front deals to establish their presence in a particular country, with the hope of reaping the rewards at a later stage through network expansion, software development and customer service activities. The proliferation of Japanese switching equipment in Asia has resulted partly from extremely competitive pricing and generous financing packages for hard pressed PTTs in developing countries. For suppliers, the really profitable business begins after the initial installation. One NEC official, in an interview in Beijing, November 1992, candidly proffered the view, 'Japanese equipment is very cheap, but the service is very expensive'.

In China, the sudden proliferation of digital switch manufacturers in 1993 prompted an all out drive to secure market share. The impact of this strategy in China has already been demonstrated in the mobile communications industry where Ericsson took advantage of the temporary inactivity of its main competitor, Motorola Inc., to build a commanding position as a supplier of cellular network equipment. A disagreement between Motorola and Chinese government officials in the 1980s temporarily froze the Illinois-based supplier out of the running for new cellular network contracts. As a result, Ericsson was able to lock Motorola out of some key markets, including Guangdong province, which has enjoyed the fastest growth in mobile services in China in recent years. Ericsson was able to do this because although cellular network equipment is built to allow different manufacturers' terminal equipment to work on different networks, the network equipment itself cannot necessarily be connected with other vendors' mobile switches or radio base stations.

China's Ministry of Posts and Telecommunications has begun a project to enable automatic subscriber roaming between users of Ericsson-supplied cellular networks in different cities. A parallel project is being carried out to link together Motorola's cellular systems in China. Once automatic roaming is possible within each of these networks, the MPT is planning a national project to link together the Ericsson and Motorola networks. However, it may be some years before this becomes a reality. Unlike analogue cellular systems, the GSM specifications provide for open interfaces between different manufacturers' network equipment. The Pearl River Delta GSM network, for example, went live in October 1994, using mobile switches and radio base stations supplied by four different vendors.

Equipment Standards and Asian Markets

Differences between vendors' equipment are exacerbated by the problems of customizing products to local market conditions. With network products, for example, suppliers may have to customize signalling software to a national standard or make special adaptations to wireless system hardware in order to conform with radio spectrum availability. Such national network specifications are a reflection of local conditions, including traffic levels and patterns of customer usage. National specifications also result from regulatory peculiarities, such as which authority has control of the radio frequencies in a particular country — the military, the telecommunications operator or the broadcasting authority. Therefore a wide variety of telecom specifications is inevitable in different markets and this is particularly the case across Asia's fragmented and diverse communications industry. For instance, customizing the software which controls digital network switches to take account of different national requirements has become a major undertaking for vendors.[16]

According to Arnold Penzius of AT&T Bell Laboratories, each country's switching system requires approximately 10 million lines of customized software code. Bell Laboratories estimates that 60 per cent of all telecommunications development costs are now eaten up in software development activities.[17] A huge effort is therefore required from vendors to support a national switching system in any one market and the problems are multiplied across international borders. In addition, customers want to buy systems which are upgradable to take advantage of the latest technological developments.

While only the major vendors have the corporate and financial resources needed to support a traditional network systems business on this scale, the same issues apply to suppliers serving the customer premises equipment market. The complexity of type approvals processes, for instance, can add significantly to the cost of marketing products in the region. A study commissioned by the Asia Pacific Telecommunity (APT, 1992) revealed that suppliers collectively spent US$140 million on type approvals in member countries during 1991. The study further showed that in order to introduce a product with a two to three year life cycle into 'multiple Asian markets', type approval procedures would account for 30 per cent of the supplier's marketing costs. It could be argued that common equipment certification is unlikely to become a reality soon because of the large sums of money Asian governments earn from type approvals and testing.

Many governments in the region, however, clearly recognize the benefits of a liberalized telecommunications equipment market and the necessity of streamlined type approval procedures. Both the Asia Pacific Telecommunity and APEC's Working Group on Telecommunications (APEC, 1994b) have addressed this issue as a priority. (For details of standards setting regulations, see also APEC, 1993, 1994a.) However, the requirement for consensus among member economies is likely to result in a long drawn-out process to simplify procedures across the region (Hawkins, 1994). The lack of a common Asian approach to standards setting is a fundamental issue affecting supplier strategies in the region and is manifested in the hotch-potch adoption of US, European and Japanese network standards. The matter is complicated by the different objectives underlying equipment certification which national telecommunications administrations may have — objectives other than the generally accepted minimum requirement that a piece of equipment should not harm the network to which it is connected. Early in 1994, for example, China's MPT introduced new certification requirements for cellular handsets in order to combat the increasing problem of fraudulent use of mobile telephones. While this has not prevented small quantities of uncertified cellular telephones still finding their way into the market, suppliers like Ericsson have faced lengthy delays in working through the certification procedures.

The cost to suppliers of carrying out these procedures may be tolerable, but combined with the prices of other modifications to equipment for international markets, they present a significant cumulative additional cost. AT&T for example, estimated it spent US$750 million on 'internationalizing' its product range for world markets in 1991. Such a huge burden of cost is a reflection of the number and size of different markets and inevitably end users in the smaller markets have to pay a higher price for products tailored to their country's requirements. Cellular network operators in Hong Kong, for example, have complained they are unable to obtain volume discounts on mobile telephones from suppliers who are more intent on serving the US or major European markets. To illustrate the point further, a mobile telephone which cost US$1,200 in Hong Kong in 1993 could be purchased for as little as US$160 in the United States, US$200 in the UK, or alternatively US$3,700 in Indonesia (Goldstein, 1994, p.14).

Cellular

The transition to digital cellular services has exacerbated the problems of equipment supply faced by operators in the region. Those who launched GSM services in 1993 had to compete for handsets against European markets such as Germany, which have measured their subscriber growth in millions rather than the tens of thousands of customers who are currently signing up for digital services in places like Hong Kong and Singapore. Cellular operators implementing the digital AMPS standard in Asia have found themselves similarly low down the international suppliers' pecking-order. Recession slowed up the deployment of digital AMPS networks in the United States, and hence the roll-out of the new generation of TDMA and dual mode handsets. This means the volume production of handsets has taken longer to materialize than originally predicted and prices to the end user have consequently taken longer to fall. However, the small size of markets like Hong Kong has been only partially responsible for the comparatively high price of cellular terminals; factors other than product volumes and development cycles affect the cost of equipment to customers. For instance, cellular service providers in the UK have long subsidized the cost of customer equipment in order to promote subscriber growth. Prices have come down significantly in Hong Kong in more recent times, primarily as a result of competition as mobile operators have fought to drive up their digital subscriber bases. Similarly, a start-up analogue network operator, Smart Information Systems in the Philippines began subsidizing handset prices in order to lower the cost to customers to around US$250 in 1994. Furthermore, major operators from the US and Europe, like McCaw Cellular Communications Inc and the Vodafone Group are able to leverage their domestic mass markets to encourage hardware suppliers to provide better service to their Asian operations.

Equipment Suppliers as Service Providers

More directly, suppliers such as Motorola actively participate in certain markets as service providers. In Hong Kong, Hutchison Telephone Company Ltd's digital cellular investment strategy has been heavily influenced by partner Motorola's global research and development strategy. The major equipment vendors clearly recognize that entering the telecommunications market as service providers presents risks. European telecommunications operators said they have been reluctant to procure AT&T switches in the past because of the US carrier's attempts to move into the European network services market. Likewise in Asia, a cellular network operator would be unlikely to procure a particular manufacturer's GSM system if the same vendor were, hypothetically speaking, an investor in a rival GSM network. Alcatel has disclosed a propensity to align itself with individual operators in certain cases. For example in Germany, Alcatel has an equity stake in a GSM network and in Hong Kong for a time, it acted as a technical partner in a GSM bid and subsequently provided technical assistance to a cable TV project which failed to get off the ground. (See also Chapter 6.)

For manufacturers, moving into the services business may offer the prospect of longer term revenues with higher returns than might be expected from equipment sales. Their participation in a services operation can also lock their co-investors into using their particular product line. Indeed, in highly competitive bidding situations, a vendor's only realistic chance of supplying a network may be through its equity participation in the winning consortium. There are also varying degrees of supplier participation. For example, providing consultancy services and a supplier finance package together with network equipment in exchange for a share of network revenues could be viewed as a manufacturer participating in the network services market. The advantage of this arm's length approach is to avoid prejudicing relationships with existing customers.[18]

The telecommunications industry's move towards open systems, specifications and standards is undoubtedly changing the relationship between rival suppliers and their customers. In the past, a cellular operator's selection of a particular analogue system locked them into one vendor because of the incompatibility of different suppliers' equipment. As discussed earlier, GSM is intended as an open standard which enables different vendors' products to be used in the same network and it is clear that network operators will increasingly have a choice of supplier which didn't exist before. This has enabled former rivals among network equipment suppliers to offer combined systems solutions. Examples are Ericsson's demonstration of the interoperability of its switching system with other suppliers' base station equipment or Northern Telecom's short-lived venture with Motorola to provide a digital AMPS cellular system. Motorola's decision in the USA to focus on CDMA technology affected its relationship with both Northern Telecom and Hutchison.[19]

Supplier Finance

These examples illustrate other strategic imperatives facing suppliers operating in world markets. Operators are increasingly demanding a choice of network equipment suppliers. At the same time, suppliers are finding it increasingly expensive and challenging to provide complete systems solutions for a proliferation of mobile and fixed network standards. The trend towards open systems and increased customer choice has certainly begun to appear in the fixed wire arena, where national carriers are now able to invoke competition between network vendors long after initial supply deals are signed. For instance, in December 1993, Telecom Australia formalized its first agreements with suppliers under its Future Mode of Operations digital upgrade programme. The agreements constitute the Australian carrier's first strategic partnerships which tie suppliers into a 'world's best practice ' and quality standard.[20] The agreements restrict suppliers to narrow cost margins and allow Telecom Australia to bring in a second supplier if one of the principals fails to perform.

Customizing products to meet different national standards imposes a higher burden of cost which suppliers pass on to their customers among the Asian governments, telecommunications operators and ultimately consumers. Of course, clearly defined national standards are part of the price which must be paid for smooth network evolution; the issue is to what extent national standards are determined by straightforward technical criteria or, alternatively, how much they are influenced by extraneous political, economic and commercial considerations. Operators who fail to adequately address these issues are storing up problems for the future:

> The present day telecom manager spends time in planning, procurement and installation of equipment, but he does not allocate enough time for acceptance testing and quality improvement. His emphasis is on providing the services as soon as possible and in doing so, he often resorts to short cut procedures. The motto here should be — do not compromise on quality; you are robbing Peter to pay Paul. The problem will become more acute when the network grows. With the growth of the network there are problems of interconnectivity, standards, type approvals, field acceptance testing, calibration and several others. In many countries of our region, these problems are not being studied seriously and there is a need to create an awareness for addressing them.
> (Former Executive Director of APT, Chao Thongma, 1992)

Several major suppliers have suggested these issues collectively pose the most pressing problem facing China's telecommunications network — which in ten years, according to current growth projections, will become the largest national telecommunication network in the region, if not the world. China's problems have been exacerbated by the country's willingness to buy different varieties of telecommunications equipment in exchange for varying sources of supplier finance and inter-governmental soft loan aid. Soft loans were particularly important for overseas telecommunications equipment suppliers seeking to export to China until the Helsinki agreement of January 1992, when the OECD countries decided that infrastructure projects which were economically viable should no longer qualify for soft loan aid. AT&T has candidly admitted its frustration at being unable to benefit from the soft loan funding enjoyed by European and Japanese suppliers to China, but has nonetheless formed an impressive consortium of international banks and set up an extensive supplier financing operation to remain competitive in the field of project finance.

Japanese equipment suppliers were equally frustrated that funds from their government's Third Yen Credit package were opened up to bids from manufacturers outside Japan.[21] In this context, it is perhaps not surprising to hear arguments that concessionary finance distorts market development. Indeed the Australian Bureau of Transport and Communications Economics stated in July 1991 that Australian companies were likely to lose out because of such distortions:

> Many exporters, including firms involved in exporting telecommunications equipment and services to the Asian market, are provided with some form of concessional finance by their governments. Concessional finance, in most cases, is a subsidy for telecommunications services and equipment and like any subsidy, distorts free market outcomes and tends to lead to misallocation of telecommunications services and equipment. They attempt to justify these subsidies in terms of development aid by making relatively cheap services and equipment available to less developed countries. However, commercial decisions to increase market penetration rather than development aid appear to be the priority in relation to the administration of their concessional finance . . . International practices in concessional finance and other distortionary policies are expected to become increasingly important for Australian firms. Even though the Australian telecommunications industry is expected to become more competitive internationally, Asian contracts may be lost by Australian firms because of distortionary trade policies implemented by governments of overseas competitors.
>
> (BTCE, 1991)

The Vietnamese government has recognized the ability of telecommunications projects to generate their own finance, and in past years has sought to channel inter-governmental loans away from the telecommunications sector into other infrastructure activities such as building hospitals, road and water supply measures.[22]

Recognizing the growing importance of alternative sources of project finance, equipment suppliers have increasingly leveraged the multinational characteristics of their organizations to tap into different sources of government loan finance. Hence a supplier like Alcatel, for example, has been exporting telephone exchanges to China from several different European subsidiaries using finance from different governments. It is therefore understandable that developing countries should welcome offers from abroad to supply top-of-the-line telecommunication equipment and attractive finance packages to pay for it, even if it means compromising on longer term technical issues, like interoperability. It may also mean compromising on price. If a government finance package effectively ties a recipient to buying equipment from a national manufacturer, the customer may have a choice of only one supplier and hence little leverage to reduce the price of the equipment. This is particularly prone to happen in countries where the government retains control over the telecommunications operator and technical criteria can fall prey to political or short term financial considerations. (See Manapat, 1993.)

National Standards

In some instances it is questionable whether government decision makers are fully conscious or even bothered by the cost implications of standards practices. For example, China's State Radio Regulatory Commission (SRRC) in 1993 recommended that non-international standard frequencies be used for CT2 networks in China. This recommendation was approved early in 1994, though it has yet to be seen whether the standard will be widely adopted or ignored by provincial telecommunications authorities. Foreign equipment suppliers had warned that such a shift in frequency standards would necessitate expensive modifications to CT2 radio hardware and software and an additional six months of product development activity. However, vendors also acknowledged that the China CT2 market was potentially big enough to financially justify making such changes to their products. The SRRC put forward no definitive reason for its recommended radio frequencies, although there has been some industry speculation that it was either politically or economically motivated.[23] Politically, it is argued that the SRRC wanted to avoid a clash with military authorities which manage the part of the radio frequency band which has become the de facto international CT2 standard. Economically, by creating CT2 products which are specifically engineered for the China market and manufactured in China, the government can maintain control over the supply of equipment. In addition, it promotes self-sufficiency in its lines of supply — a political imperative for China since the events in Tiananmen Square in 1989 which prompted an international moratorium on new soft loan funded telecommunication development projects.

Vendors contend that Japan has used its own unique national telecommunications standards as a means of restricting market access by foreign suppliers. For example, the US government and Motorola have been lobbying for better access to the Japanese cellular market since the mid 1980s. The matter was escalated to the level of a serious international trade dispute in early 1994, when the US government threatened to impose punitive tariffs on Japanese telecommunications equipment imports unless Motorola gained better market access to Japan. The argument related to Japan's development and imposition of its own cellular standards, which the US has long contended constituted a barrier to market entry for foreign equipment suppliers. Finally in 1989, the Japanese government bowed to US government pressure and allocated frequencies for the implementation of the more widely used TACS cellular standard in order for US suppliers to market their products in the country. However, the designated TACS operator, Nippon Idou Tsushin (IDO) had already begun implementing its own Japanese cellular network and dragged its feet in building out the new TACS network. Such was the delay that the US government threatened to invoke Super 301 sanctions against Japan in 1994. US trade negotiators eventually brokered an agreement which involved IDO placing orders worth US$300 million for Motorola network equipment to be installed by the end of 1995.[24] It is worth noting that since the start of its dispute with Japan, Motorola set about manufacturing cellular telephones conforming with Japanese standards for the Japanese market.

Towards Common Standards

Having unique national standards is a double-edged sword for Japanese manufacturers. They have been notably unsuccessful in exporting the Japanese Digital Cellular standard to other Asian countries. The Japanese Telecommunications Industry Federation is now hoping for greater success in getting its Personal HandyPhone system (PHS) — a digital cordless access system — adopted by other Asian countries.[25] The adoption of unique national standards by NTT has also posed a dilemma in export markets for Japan's switching industry. Companies like NEC have developed special versions of their products to serve international markets since the indigenous system is designed to different technical specifications, in the same way that Europe and the United States employ different network channelling configurations.

Some commentators contend that in the long run, such technical isolation could undermine the long term viability of the Japanese switching industry. While export versions of NEC and Fujitsu public switches have enjoyed some success in Asia, it has been argued that these suppliers will gradually be overtaken by European and American manufacturers. Producing a switch for international markets is a problematic and expensive business. For instance, NEC's attempts to localize production of its NEAX system in Tianjin, China, were delayed because of difficulties in meeting the MPT's national specifications for ISDN working as well as COCOM[26] restrictions. Some experts predicted that the widespread deployment of PHS outside Japan could be inhibited by the necessity of producing new switching software code for the various systems in use overseas.

Another difficulty facing Japanese suppliers is the country's reluctance or inability to promote its own technical standards in global forums. Dr Lance Wu of Taiwan's Industrial Technology Research Institute argues that Asian countries outside Japan are involved in the periphery of standards setting and adds that even Japan provides little regional leadership in international standards fora:

> Asian countries are doing very little on the standard setting side and there are various reasons. Asian countries are very much followers of technology. In Taiwan we have a very strong PC industry, but we are not setting standards. That is done by Intel and Microsoft. I don't think there is going to be any change in the [near] future for Asian countries to participate more actively in setting standards. Another issue is the language issue. Standards meetings are often political as well as technical arenas. Your representatives have not only got to be able to talk in English, they have to be able to fight in English.
> (Dr Lance Wu, Deputy Director Computer & Communication Research Laboratories, Industrial Technology Research Institute, in an interview with the author, December 1993.)

The ITU has also recognized that developing countries, not just those in Asia, have legitimate concerns to air in upstream standards activities and has set up a special study group to investigate ways of enhancing developing nations input into standards setting:

> Today certain developing countries do not have the resources to adequately document, represent or promote their special concerns (such as simplicity, robustness, environmental tolerance, modularity etc.) in standard setting

processes, but they, themselves are members of various regional and global organisations which participate in ITU standardisation activities.

(ITU 94/6 *Experts to Consider Recommended Approaches for Harmonisation of Networks*)

In reality, however, the current trend in Asia towards coordination of standards activity often appears to be one of confrontation rather than cooperation. For example, Japan's promotion of its own version of ISDN in the region has been criticized by Australian authorities and failed to make inroads into world markets. Some suppliers argue that historically speaking, Japanese manufacturers established a foothold in Asian markets such as China and Thailand at a time when US and European manufacturers were unwilling to transfer their technology to Asia or were more focused on selling to their own domestic markets.

Suppliers' Shift to Asian Strategy

In summary, therefore, it seems reasonable that local market conditions comprise only part of the combination of influences which dictate the various strategies of telecommunications equipment suppliers operating overseas. Indeed, the evidence indicates that domestic influences in traditional home markets have an immense impact on supplier strategy overseas.[27] Table 7.6 lists the activities of the main equipment suppliers.

Northern Telecom, as a pure equipment supplier, has sought to diversify from its past monogamous dependence on the North American digital switch market. Its concerted efforts to break into the Japanese market, discussed above, could be interpreted as an attempt to recreate its success in the sophisticated telecommunications market of the United States. More recently, the company decided to focus on developing its business in China and this constitutes another significant shift in strategy, which Northern Telecom president Jean Monty acknowledged in Beijing in November 1993. Monty said that China had become the company's single most important international market.

AT&T chairman Bob Allen expressed similar views about China after the company signed its memorandum of understanding with the State Planning Commission in February 1993 and he subsequently predicted that China would remain the world's single biggest telecommunications infrastructure market for the next 30 years.

This shift of focus to Asia is reflected in the geographic shift of revenues which suppliers are anticipating.[28] Northern Telecom, for example, estimated the Asia-Pacific region contributed 7 per cent of its global revenues in 1993. A 50 per cent increase in sales in the Asia-Pacific during 1994 was expected to increase the region's share of worldwide revenues to 10 per cent. Alcatel reported that 10 per cent of its global revenues were generated in the Asia Pacific in 1992, a figure it hopes to increase to 20 per cent by the turn of the century. Executives say the company is currently making between 10 per cent and 20 per cent of its global sales in the Asia-Pacific region. In 1991 approximately 17 per cent of AT&T's revenues — including telephone calls — came from international markets. The company intends to increase this to 25 per cent by the mid 1990s and to 50 per cent by the end of the decade. Currently less than 10 per cent of the company's US$11 billion equipment sales are to the Asia-Pacific region.

Ericsson says that Asia generated 20 per cent of its sales to the third quarter of 1994, up from 15.5 per cent during the same period in 1993. The company hopes to increase its Asia sales contribution to 25 per cent by 1996. Siemens AG reported its Asia business doubled in size in 1993 and now accounts for 8 per cent of its worldwide revenues. Motorola Inc estimated that 26 per cent of its revenues were generated in Asia in 1993, including sales in Japan, China and the Pacific rim countries. Among the Japanese vendors, NEC estimated approximately 25 per cent of its telecommunications equipment sales are generated in overseas markets.

Network Management

The accelerated development of Asia's telecommunications infrastructure has generated a requirement not just for hardware but also for the carrier expertise needed to plan and manage modern networks.[29] Closely tied to this is the development of the customized software needed to manage those networks. The supply of such expertise is generally limited to existing operators of modern networks, so telecommunication monopolies in Asia, particularly those which have adopted a fast-track development of their network infrastructure have no alternative but to import such skills from abroad in the form of a technical cooperation agreements with foreign carriers, consultancy contracts or direct equity participation. The major equipment suppliers operating in Asia acknowledge the necessity of this type of carrier involvement in the development of green field site projects such as Build-Transfer-Operate overlay network schemes as well as in any significant network expansion. Indeed, some manufacturers concede that simply having a financier and an experienced network equipment supplier is not enough to implement are extensive modern network.[30] This reflects both the increased complexity of network equipment available today as well as the need to manage effectively any form of rapid network expansion, especially that involving a multiplicity of different manufacturers' equipment. The

Table 7.6
Regional Presence — Asia's Main Telecom Equipment Suppliers

	Alcatel	Siemens	AT&T	Northern Telecom	LM Ericsson	Motorola	NEC	Fujitsu
China	GSM Transmission/m ICs/m Switching/m	GSM/m Switching/m Transmission/m	PABX/m, ICs/m VSAT, AMPS Cable & fibre/m Transmission Switching/m R & D	GSM, R & D PABX/m, ICs/m Data Switches	AMPS/TDMA GSM, TACS Switching MSC/m Pagers/m	CT2, TACS Pagers/m GSM	ICs/m Transmission Fibre/m Switching/m Pagers/m	Transmission/m Switching/m
Hong Kong			CPE/m		AMPS, TDMA GSM	CDMA, CT2 TACS, AMPS Mobile Data	SES Switching	Transmission Switching
India	Switching/m	Switching/m	Switching/m Transmission/m CPE		Switching Fibre/m	Pagers/m	SES	Transmission/m Switching/m
Indonesia	Transmission GSM	Transmission Switching/m	Switching/m	PABX CPE	PABX, GSM Transmission NMT 450	Pagers	Switching/m	Transmission/m
Japan			Fibre/m JDC	R & D Switching	PDC	TACS PDC Pagers	ICs/m, CPE/m Switching/m Transmission/m Pagers/m Fibre/m	ICs/m Transmission/m Switching/m Fibre/m
South Korea	Transmission		Transmission ICs, Fibre/m AMPS Switching	Transmission CT2		AMPS Pagers/m		Transmission/m

Asia — The Supplier's Dilemma 221

Table 7.6
(Continued)

	Alcatel	Siemens	AT&T	Northern Telecom	LM Ericsson	Motorola	NEC	Fujitsu
Malaysia	Switching/m R & D		ICs/m	CPE/m	WLL, TACS NMT 450 Switching/m AMPS/TDMA	Pagers CT2 Mobile data	Pagers DAX, Fibre Transmission/m Switching/m	Transmission Switching/m
Philippines	Transmission	CPE/m, GSM Components/m Cable Switching	AMPS, TDMA Switching	Components/m	TACS Outside plant Fibre	N-AMPS AMPS	Transmission/m Switching	
Singapore	Switching	ICs/m	ICs/m CPE/m	PABX	Mobile data ETACS GSM	Mobile data Pagers/m	SES, PABX Transmission	Switching
Taiwan	SES Switching/m Transmission/m	Switching/m	Fibre Switching/m	Data Switches GSM	AMPS		CPE/m	Transmission/m
Thailand	Switching Transmission	Switching/m	Transmission ICs/m, CPE/m Switching/m	PABX, CPE Components	Outside plant/m AMPS, GSM Transmission Switching/m PABX	Mobile data Pagers AMPS	Transmission Switching/m CPE/m	
Vietnam	Transmission Switching/m GSM	Transmission Switching/m Fibre/m		GSM Switching	GSM AMPS/TDMA Switching		Switching Transmission	
Region	Submarine Systems		Submarine Systems				Submarine Systems	Submarine Systems

Note: 'm' denotes manufacturing facility.
Source: Supplier data.

ITU's Telecommunication Development Bureau (BTD) addressed this issue and related concerns over standards of basic network maintenance in developing countries at the Regional Telecommunication Development Conference in Singapore in May 1993:

> While it is true that most of the countries of the region have been able to cope quite well with the maintenance of high technology, it is also unfortunately true that the network services standards for both quality and reliability are inadequate. There remains throughout the region a poor call completion rate and there is a persistent 'last mile' problem. Procedures and practices and installation and maintenance techniques for the low technology cable reticulation remain poor. These problems have been addressed. . . . Much remains to be done however and further support has been sought to introduce modern network management techniques to the networks of many of the countries.
>
> (ITU BTD *Document 31* AS-RDC/93 8/5/93)

Most countries in the region have liberalized the supply of telecommunications equipment, but overseas involvement in the management and operation of networks remains a sensitive issue. Nearly all Asian countries maintain strict limitations on the level of foreign involvement in managing the national telecommunications infrastructure. Therefore, direct equity participation, or foreign direct investment, by overseas carriers is the exception rather than the rule. There are some notable exceptions to such controls on foreign enterprise (such as New Zealand and Australia) and notable examples where overseas equity participation is prohibited or monopoly control has been rigorously maintained, as in China and Singapore. Faced with these realities, foreign carriers with experience of operating sophisticated modern networks have found a ready market for their services, but in Asia particularly, the regulatory framework for their participation is highly restrictive.

The major carriers certainly recognize the value of their network operations expertise and are not prepared to give it away, especially if more money can be made by employing their experts in the home country rather then loaning them out to other carriers.

With the growing rivalry between the major carriers operating in international markets, consultancy skills are also increasingly seen as a strategic commodity which should not be loaned out to potential competitors. So in general terms, telecommunications consultancy work is viewed as a comparatively low margin proposition by major carriers, and is often approached as a pump-primer or quid pro quo for more lucrative network management and turnkey contracts or alternatively as a basis for taking equity in a country's telecommunications network. An associated trend is the growing necessity for developing countries to improve their standards of telecommunications management. In the increasingly competitive market for sources of international finance, these developing countries require consultancy expertise to put their operations on to a more business-like footing in order to satisfy the requirements of international lenders such as the World Bank, the Asian Development Bank, foreign governments and commercial lenders.

In Asia, the ability of overseas carriers to export their skills has depended greatly on the individual regulatory environment in each country as well as each carrier's strategic objectives in how it executes its international business. Nynex Network Systems Company, for example, the overseas operating arm of the US regional Bell company, Nynex Corporation, has sought infrastructure development opportunities in Asia which allow it to leverage its expertise as a telecommunications carrier along with its technology 'vision' of integrating cable TV and telephony services on a common network. This can be seen in Thailand where Nynex has a 13.5 per cent stake in TelecomAsia Corporation, which is building a two million line overlay telephone system in Bangkok. The project is due for completion in 1997 and TelecomAsia has planned to deliver cable TV services starting in 1995 using the same telephone network but operating under a separate franchise agreement with the national broadcasting authority, the Mass Communications Organization of Thailand. (See also Chapter 6.) Nynex's previous experience in this field has included investing US$3 billion building out an integrated cable TV and telephony network serving 2.7 million homes in the United Kingdom through its 19 Nynex Cable Comms Ltd franchise areas. Senior Nynex executives stated that the most sobering lesson from their UK experience has been the necessity of sourcing entertainment software, rather than any technical or marketing issues. When the company began its move into the UK cable TV, it found many of the TV distribution rights for some years ahead had already been purchased by the direct to home satellite broadcaster BSkyB.

Nynex's US$1.2 billion investment in Viacom in 1994 and the subsequent US$10 billion acquisition of Paramount Communications is a direct reflection of this strategic imperative. The relevance of this issue has yet to be fully addressed by Asian carriers, many of which are considering providing cable TV or video-on-demand services. If Nynex's UK experience is relevant in Asia, the biggest challenge will not be obtaining suitable broadband switching equipment, video servers or software algorithms, but rather in obtaining programming rights.

As a measure of Nynex's strategic aspirations in terms of developing cable telephony networks, the US carrier retreated from a cable TV and telephony project in Japan early in 1994 involving the Tomen trading company and the Yokahama cable TV operator, because at that time, the Ministry of Posts and Telecommunications in Japan would not permit the group to offer bundled cable TV and telephony services. In Bangkok, meanwhile, Nynex estimated the incremental cost of upgrading the TelecomAsia network to carry cable TV signals at US$250 per subscriber line, in addition to the US$1,000 cost per line of the initial telephone connection. Most of the additional cost would be attributable to the backbone fibre infrastructure and headend equipment rather than the local drop. Following the initial public offering of TelecomAsia shares in December 1993, Nynex estimated it had quintupled its original investment in the company in under two years and has been attempting to extend its Thailand model to other Asian countries, including China.

It is worth recalling that Nynex's financial success in Thailand was not achieved without a measure of risk. Indeed, BT (formerly British Telecom) was Charoen Pokphand's original joint venture partner, before the UK operator reduced its role to that of a technical consultancy. BT's withdrawal came at a time of political turmoil in Thailand with the reassertion of a military government and the ruling junta's decision to re-examine the terms of CP Group's original franchise as part of a drive against corruption and favouritism. (See also Chapter 3.) At the same time, BT was re-evaluating its globalization strategy and withdrawing from overseas infrastructure projects in order to focus on its network services business for corporate customers. The company also decided that absence of any significant political risk should be a key prerequisite for future overseas investments. BT had meanwhile decided to wind down its activities as an overseas equipment manufacturer, in particular to divest itself of its Canadian subsidiary Mitel, which produces rural and private branch exchanges.

Signposts for the Future

The TelecomAsia project in Thailand highlights an emerging technology imperative for telecommunications equipment suppliers — that Asian operators are increasingly looking to achieve more than providing a plain old telephone service. The ability to deliver, in the first instance, television entertainment through the telephone infrastructure and subsequently a range of interactive multi-media services, holds out the prospect not only of greater revenues but also reduced incremental costs of building out duplicate cable TV and telephony infrastructure.[31] Telecommunications planners at the Hong Kong-based Wharf Holdings group have hypothesized that China could shave up to US$30 billion from its projected spending on communications infrastructure by the year 2000 by building integrated cable telephony networks. Wharf itself is investing an aggregate US$1.4 billion in a full service network in Hong Kong and is engaged in several communications projects in mainland China. (For the philosophy behind this, see Lam, 1993.) Some incumbent telecommunications operators are looking to speed up the provision of basic services using the installed cable TV infrastructure. (See Chapter 6, Table 6.1). While this holds out significant market potential for equipment suppliers, the regulatory conflicts between broadcasting and telecommunications authorties in many Asian countries has not provided encouraging signs of a smooth convergence of communications technologies.

Broadband in Asia

While Asia's information superhighway is likely to remain a populist piped dream for years to come, carriers are actively researching technologies which will support broadband services. This, in turn, has impacted on the suppliers' strategic approach to product development and marketing in the Asia-Pacific area. Alcatel's solution for converting TDM digital switches to ATM working which was referred to earlier in this chapter, is one example of this. They, along with other suppliers, are increasingly being asked by their customers to assist in projects to develop broadband network products in Asia. In Taiwan, the Telecommunications Laboratories of the Directorate General of Telecommunications asked its three national switch suppliers to cooperate in an indigenous programme to develop a central office ATM switch.

This posed an interesting dilemma for joint venture suppliers Alcatel, AT&T and Siemens, which had already invested years of effort and billions of dollars developing their own ATM systems to sell to customers like the DGT. In theory at least, they were subsequently being asked to assist in the development of a competitive product line. With the public network switch market approaching saturation in Taiwan, suppliers led by AT&T are diversifying into the cable TV business, providing turnkey systems, consultancy and equipment. Hence AT&T is able to supply two different sections of the communications industry in the expectation that market liberalization will ultimately enable it to sell the same telephony and video-on-demand products to either the cable operators or the DGT, which are likely to be competing with one another.

Both these examples reflect the constant shift in the relationships between multinational suppliers and their customers. Localization has clearly come to mean more than the assembly of semi-knockdown telephone exchanges with the addition of a quota of locally sourced materials. Multinational suppliers are having to weigh-up and balance the often competitive aspirations of customers in their various regional markets. Taiwan's aspirations to develop its own central office ATM switch are all the more remarkable because the island has no previous experience of public network switch development. A previous attempt to produce a digital PABX failed to become a commercial reality because the system was too large and, according to local experts, it missed its 'time window of opportunity' for export markets. Even those responsible for the ATM project have compared their ambitions to 'trying to organise a moon landing.'(Wang Jin-tuu, managing director of Telecommunication Laboratories, described the challenge these terms in an interview in December 1993.) The DGT intends that its ATM product should serve both the export and the domestic market.

In South Korea, the government launched a national effort to develop a broadband network infrastructure, including an ATM version of the locally produced TDX switch. The Highly Advanced National (HAN) communications project has been funded partly by the sale of shares in the national carrier, Korea Telecom. The TDX switch has made only small inroads into telecommunications markets overseas and its export successes to Russia and the Middle East have been heavily supported by Korean government concessionary finance. Hence planned development of an ATM version of its switch and a range of SONET transmission equipment is a fairly ambitious step. (See *Communications Week International*, 6 September 1993.) In this respect it is worth noting that Asian countries appear to be becoming more ambitious in their development of telecommunications network products and in the level of technology which governments and manufacturers are prepared to invest in.

Local Manufacturing

Until now, much of the local enterprise in telecommunications products has focused on original equipment manufacturer business, leveraging the low labour costs in Asian countries. As markets have matured, local manufacturers have transferred their OEM (Orginal Equipment Manufacturer) business to lower cost countries and attempted to develop higher value products in their home markets. Several of the major Hong Kong based manufactures operating in China have followed this path, including Champion Technology (radiopaging equipment), Double Kingdom (cordless telephones) and S Megga Telecommunications (customer premises equipment).

The cost structure of Asian telecommunications manufacturers can be extremely competitive and is closely attuned to local market requirements.

In Taiwan, for example, the DGT's purchase price for a basic telephone set for connection to a local line may be as low as US$10. Vendors such as AT&T argue they cannot compete at such low profit margins. Significantly, Taiwanese manufacturers are also able to export the same low cost telephone sets to Malaysia, where they can undercut local suppliers such as Sapura Holdings. Sapura, in turn, says it targets west European markets such as Germany where its products are of a sufficiently high value, but can be priced cheaper than equivalent European-produced goods.

Research and Development

The high cost of core research and network product development has led Asian manufacturers to concentrate on niche market products — Asian language radiopaging devices are a good example — customer premises equipment and accessories ranging from low-tech devices such as tone diallers to digital cellular telephones. These products collectively make up a small but significant share (approximately 10 per cent) of the total telecommunications equipment market. At the other end of the scale, Asian countries which are investing heavily in core telecommunications research for national strategic reasons often have difficulty marrying up their technology with entrepreneurial expertise. Suppliers report this is particularly the case in China which has built up some formidable telecommunications research and development expertise in recent years, but has been unable to commercialize its technical resources. With the winding down of COCOM and an end to its technological isolation, China has increasingly looked to overseas suppliers to capitalize on its technical advances. Early in 1994, for example, the Ministry of Posts and Telecommunications began searching for a joint venture supplier to help develop and manufacture a Chinese 2.4 Gbit/s synchronous digital hierarchy or SDH transmission system.

In this instance, overseas expertise was needed to develop the network management and signalling software for the new product range, rather than the core system which has been developed in China. A similar requirement for foreign expertise and cash can be found in two TDMA digital network switching systems, the DS30 and the HJD04, which have been developed by engineers in China, but which for several years failed to bridge the leap from the laboratory to the production line.[32]

The same issues are likely to apply to the results of a series of ATM switching projects underway in China's research institutes and universities. George Smyth, president of Bell Northern Research, said his company's single greatest contribution to China's telecommunications development may ultimately be in the sphere of commercializing products rather than in core research and development.

ATM is quite well publicized and the technique is quite well understood in China. I would say the Chinese fully understand ATM's capabilities and they know where to apply it. I think the difficulty is China's ability to commercialize its technology and the rate at which they do this — the way they organize the flow of technology to the implementation and manufacturing is the [main] problem. There are many difficulties. One difficulty is the rapid growth of the network itself. No one has experienced or tried such a rapid growth before. There is also a rapid evolution in technology. The technology is being updated and outdated very quickly. There is also a rapid evolution in regulation . . . a state of creative chaos.

(George Smyth, BNR, Beijing, November 1993, in an interview with the author.)

Conclusion

Strategically, therefore, equipment suppliers have had to move away from shipping boxes and instead, are transferring technologies and skills which correlate with indigenous technology developments. There remains, of course, the commercial imperative for such ventures to provide suppliers with a sufficient return on investment. This is weighed against the size and potential of individual markets and the strategic necessity for suppliers of establishing a presence in those particular countries. For example, Ericsson's plans to set up a research and development facility in Taiwan have been deferred because of the company's inability to secure access to the public switch market. This prevents Ericsson from achieving the level of ongoing business required to sustain such a research operation, despite the fact that it is the DGT's primary supplier of cellular equipment. For its part, the DGT, like its counterpart in mainland China, does not want to burden itself with the technical complexities of introducing additional switch suppliers into its market. However, Taiwan's aspirations to join the World Trade Organization (WTO) may mean that the Executive Yuan's moratorium on new switching systems may have to be discarded in 1995.

Experience has shown there are significant secondary benefits for multinational suppliers prepared to localize their operations abroad. Suppliers with comparatively small 'home' markets such as Ericsson have increasingly looked to their overseas subsidiaries to provide future generations of engineers and researchers. India and Malaysia have developed a considerable skills base in software engineering. The comparatively low cost of software expertise in India has proved extremely attractive to multinational suppliers and carriers themselves. Indeed, India's National Commissioner for Telecommunications in 1993, H.P. Wagle, argued the country's software expertise will assure the future of the indigenously produced public network switch, developed at the Centre for the Development of Telematics.[33]

However, it is becoming increasingly evident that with few exceptions, no one country or single manufacturer can continue to provide single-handed all the resources and research expertise needed to support the network solutions of the future. This is illustrated by the spate of alliances between switch manufacturers and computer networking companies in order to develop common open architectures between public and private networks. In the cellular telephone industry, suppliers have been working together — not always successfully — to develop network solutions which comply with the proliferation of technical standards. No one supplier has been able to provide products which serve all the current cellular standards, but equally the major manufacturers cannot afford simply to focus on one particular technology or standard. At the same time, the move to open systems is enabling suppliers to work together on common system solutions and is providing niche market opportunities for new players, particularly among the agile and highly specialized technology suppliers.

The strategic significance of this opportunity for Asian telecommunications equipment suppliers was addressed by Australia's Bureau of Transport and Communications Economics:

> It is not likely that Australia will become a competitive supplier of central switching equipment . . . because of the massive developmental costs involved. PA Consulting has reportedly estimated that 8 per cent of the world market is required before a workable return can be obtained. However, the trend toward an open network framework for telecommunication systems may reduce the economies of scale and be advantageous for Australian exporters. An open network framework allows individual components (such as software packages) to be appraised and added to the existing system. (BTCE 1991)

The competitive nature of the industry means rapid product development cycles have become an imperative issue in determining long-term success or failure. This need for speed has reached the stage that it has outpaced suppliers' ability to organically grow their organizations to meet their requirements. Mergers and acquisitions represent perhaps just a taste of things to come in a broader shake-out of the telecommunications equipment industry which lies ahead. In this context, the major suppliers' ability to tap the skills and resources of countries in the Asia Pacific region — in effect a reversal of the flows of supply and demand of the past — has become a critical strategic issue. Indeed, their long-term survival may hinge upon it.

---

[1] The Asia-Pacific region defined by the ITU includes Japan and Oceania but excludes the former USSR. Asia, in this reference, excludes Oceania. The data is based on the region's 12 biggest telecom markets. Mobile market data source: CIT Research Ltd. 1994, London.

[2] See, for example, Walther Richter, ITU Telecommunication Development Bureau, May 1993. AS-RDC/93 Economic Justification for Telecommunications Investment in Developing Countries, ITU, Geneva, Switzerland.

[3] For example, NTT spent US$2.4 billion on research and development in 1992, a 60 per cent increase on its 1988 research bill of US$1.5 billion. Lars Ramqvist, president of LM Ericsson, said his company's annual R&D expenditure is running at US$2 billion. 'We will continue to spend the money. The gamble is to expand the business,' he told the author in an interview at Singapore Telecom, May 1993.

[4] The total cost per access line in China dropped from US$1,024 in 1991 to US$850 in 1993, according to Xu (1994), and may be lower than that, according to Ure (1994b).

[5] The termination of new soft loans for telecom projects by OECD countries, tougher competition between suppliers, the opening of new markets and the increasing demands on capital have highlighted the impact of telecommunications on developing countries' debt burden. 'While developing economies benefit from the downward trend in equipment costs, the need for investment at world prices will mean that a significant expansion of the telecom infrastructure will have a disproportional impact as a proportion of GDP and or national debt' (Erbetta,1994).

[6] Manapat describes the PLDT's X-4 expansion programme thus: 'The contract between Siemens and PLDT was clearly disadvantageous to the country. The German firm got to dump its obsolete equipment on the Philippines and succeeded in making PLDT a captive market of its outmoded technology, while Marcos ... profited immensely from the project. But the country is stuck with a multimillion dollar debt and an antiquated telephone system.' Other shortcomings of 'technology leapfrog' include uneven development of human resources. 'One of the mistakes the developing countries are making in leapfrogging to advanced technologies is in the distribution of the skills base. You don't get a steady evolution of skills throughout the industry. Rather you get a few highly qualified individuals while the rest remain with a fairly low level of skills.' Julian Ehrlich, The Yankee Group Asia Pacific, in an interview, November 1994.

7 'Manufacturers turn to large users as a stable long-term market. In the United States, virtually all capital investment in equipment in 1975 was by the carriers, but in 1986, the figure fell to only two-thirds. Non-carriers bought PBXs multiplexers, concentrators, network management systems, satellite and microwave facilities' (Noam, 1994, p.26). Note also supplier projections for user equipment in the Asia Pacific region. Northern Telecom estimated the Asia Pacific market for PABX systems was worth US$800 million in 1992 and was set to enjoy CAGR up to 18 per cent CAGR until 2002 (*Asia Pacific Telecommunications*, July/August 1992). Also Northern Business Information (1993) estimated the number of PABX, keyline and centrex lines serving the Asia Pacific region should top 8.45 million by 1995, an increase of 36 per cent on the figure for 1990.

8 Interview with James Long, President of Northern Telecom Asia Pacific, in Hong Kong, November 1992.

9 In a non-attributable background interview, one senior AT&T executive commented: 'We chose not to enter the beeper market in China and if we had our time over again, we would have said it is a good place to start in the mobile market.' New Jersey, USA, November 1993.

10 Politicians from both countries lobbied hard for their respective suppliers, NEC and AT&T. Locally, strong political influence was wielded by Indonesian partners who were affiliated with various members of President Suharto's family. This example illustrates how a seemingly straightforward choice of telecom products can hinge less on technological or economic considerations, and more on political and vested financial interests. See *Far Eastern Economic Review*, 24 January 1991.

11 *The Asian Wall Street Journal* detailed the key events, background and innuendo of the saga in its reports of 12 March and 15 October 1992.

12 Fujitsu already has an agreement for the manufacture under licence of its Fetex 150 switch by the Shanghai P&T Industry Corp which produces 500,000 lines each year. Half of this production run is re-exported to China through Japan and now Fujitsu has sought to establish its own joint venture switch production facilities in China. Despite the modification to Directive 56, the MPT has resisted moves by other overseas switch suppliers to start manufacturing in China. In January 1994, it warned vendors it wished to cut down its sources of equipment to three strategic suppliers, although there has been no signs of this policy being implemented in the short term.

13 Larry N Lehmen, chief technical officer for Brooks Telecommunications Corp, argued at the *AIC Cable & Telephony Conference* in Hong Kong, 2 November 1994, that the best place to build broadband networks is in places without any embedded telecom infrastructure at all. Why spend money to build a broadband network when existing narrowband technology can provide 90 per cent of the services currently required, he argued.

14 The liberalization of equipment supply is becoming increasingly important as a means of reducing telecom infrastructure costs in Asia. This is self evident from

the benefits of competitive bidding for network equipment projects as well as in areas such as the supply of cellular terminals. This point was argued at the 9th APEC Working Group Meeting on Telecommunications held in Hong Kong, March 1994, by Keji Saga, Professor of International Telecommunications, Asia University, Japan. See Saga (1994).

[15] 'Setting up shop in several countries simultaneously makes it difficult to have a centralised production strategy and with production and assembly output duplicated in several markets, each factory is dependent on the parent company for technology such as microprocessors.' Ross O'Brien, quoted in *Asia Pacific Telecommunications* August 1993.

[16] A notable recent example is China's adoption of non-international standard frequencies for CT2. See *Pacific Rim Telecommunications*, 15 September 1993 (Probe Research Inc., New Jersey, USA). Also, see ITU 1993 AS-RDC/93 Contribution of USA Delegation to Regional Development Conference, Singapore. The primary recommendation was for an: 'Open transparent, industry-sponsored standards-setting process coupled with a nondiscriminatory certification/type approval regime to achieve a competitive and robust telecommunications equipment market and a highly advanced telecommunications infrastructure.'

[17] Penzius argues that to stay ahead in such a market is a costly business: 'Some large ventures have taken billion dollar losses, some have gone out of business entirely. It is not that these companies are naive.' (Arnold Penzius, in a presentation at Bell Laboratories, November 1993.)

[18] Siemens' relationship as a supplier of GSM equipment to both Liantong and the MPT in China has many of these characteristics.

[19] Motorola's domestic preoccupation with CDMA technology as its standard of choice for the US cellular market reduced its interest in the IS-54 standard. As the ill-fated joint venture with Northern Telecom illustrates, this has had a significant impact on Motorola's international business strategy. It seems that this technology strategy has contributed to the scaling back of its exclusive supplier relationship in Hong Kong with Hutchison Telecommunications (the holding company of Hutchison Telephone Company, Hutchison Paging, and Hutchison Communications which operate cellular, paging and CT2, and the new fixed-wire line networks respectively). Hutchison was not able to implement a commercial TDMA network, and had to change its plans to implementing narrowband AMPS and CDMA technology. See 'Hutchison's Spectrum Hopping' in *Pacific Rim Telecommunications*, 15 July 1994 (Probe Research Inc., New Jersey, USA).

[20] See for background *Pacific Rim Telecommunications*, 15 December 1993 (Probe Research, New Jersey, USA).

[21] The Japanese Third Yen Credit loan to China was a notable exception to other soft loans, since it allowed non-Japanese companies to bid for telecom projects financed by the Japanese government. NEC objected to the unfairness of this arrangement since it was not able to tender for other countries' loan funded projects.

22   This point was raised by Diana Sharpe as an adjunct to Gerald Wakefield's paper 'Changes in Telecom Regulatory Structure in Vietnam' at the Pacific Telecommunications Conference, Hawaii 1994.

23   The SRRC decision may in fact contribute to the demise of CT2 in China. Supplier estimates of China's CT2 subscriber base in October 1994, ranged from 80,000 to 100,000 (GPT) and 100,000 to 150,000 (Motorola) indicating that user growth virtually stagnated over the preceding six months.

24   In this context, it is worth noting the comment by Andrew Pollack in Tokyo for the Malaysian *Business Times*, 16 March 1994: 'Indeed, what has shocked some Japanese about the cellular phone controversy is how much effort Washington has exerted for what appears here to be the narrow interests of a single company.'

25   CSL, the mobile telephone operator owned by Hongkong Telecom, has announced plans to operate a PHS system. See *Communications Week International*, 12 September 1994, p.20.

26   Co-ordinating Committee on Multilateral Export Controls. Since 1994 all high-tech exports restrictions on China have been lifted.

27   Regional Bell Operating Companies, including Nynex and Bell South have claimed in the past that their decision to invest in overseas markets is a direct response to regulatory restrictions on their activities in the US. Rather than re-investing in their domestic networks, they are able to gain a better return on capital in high growth overseas markets.

28   Dr Hans Baur, executive vice president, Siemens AG: 'The slower pace of economic development which has affected the telecommunications industry as well, is expected to dampen growth to an average of not more than 3 per cent in the next few years in Europe and America. In view of this situation, the industry's concentration on East Asia is hardly surprising: All industry experts agree that this market will grow at more than 10 per cent annually.' Press conference at ITU Asia Telecom, 18 May 1993, Singapore.

29   For example, although China's network has expanded rapidly, congestion has got worse. Long distance circuits in south and central China were so congested in early 1994 that call completion rates were down to 15 per cent. See. *Hongkong Telecom — Banishing the Bears, Hong Kong*, Salomon Brothers, 1994, Hong Kong.

30   Antoine Auquier, senior vice president, Alcatel Trade International, among others, made the point at an Financial Times Telecommunications conference in Hong Kong in February 1994.

31   Gary Davis provided a good analysis of the incremental costs of telephony on cable systems in his paper: 'The Contribution of Cable Telephony to Revenue' to the AIC Cable & Telephony Conference, Hong Kong, October 1994.

32   The MPT's manufacturing arm also now claims to produce 1.6 million lines of the HJD-04 digital main exchange, developed jointly by the People's Liberation Army and the Posts and Telecommunications Industry Corp, according to PTIC sources.

33   'Seventy per cent of the switching technology is in software development. India

has a decided advantage over many other countries in the development of software and the whole concept of having a large basic manufacturing industry seems to have changed. We will still be here in the year 2000.' H.P. Wagle, 1993, in an interview at Asia Telecom 1993, Singapore.

8 Conclusion

The priority now being assigned to telecommunications across developing Southeast and East Asian economies testifies to the changing role of telecommunications in modern economic growth. Telecommunications is often referred to as the 'highway'— or the 'information superhighway' in the case of high-speed broadband networks— along which data travels, and is thus seen as the vital underpinning of the information economy in the electronic age. However, this imagery misses the point because it is insufficient to capture the truly radical nature of the shift in the relationship between telecommunications and the generation of value in the chain of production and services. A modern telecommunication network is a system of switches, transmission medium and intelligent functions, and it is these functions which, directly or indirectly, enter the productive process. The exposure of developing countries, even the poorest, to regional and world markets makes the building of a modern telecommunications network, at least in the major cities and trading regions, an imperative of modernization.

The evidence provided by the chapters in this book supports the view that Asia's regional development is more complex than simply a case of being a large and growing market which is attracting multinational telecommunications companies like moths in the night around a lamp. The chapters support two principal points. First, while across the region the shift towards a liberalization of the telecommunications sector and a rapid expansion of networks is driven by a common exposure to regional and world markets, the policies and trajectory — that is, the pace and direction — in any one Asian economy is the outcome of local mediating factors, such as political institutions, the role of powerful families, the role of the military, the strategic position of the economy within the region, and so on. No understanding of the dynamics of local development is possible which abstracts from these factors.

Second, local Asian capital and local Asian companies are being nurtured and encouraged by Asian governments to meet the challenge of the multinational corporations, be they common carriers, regional broadcasters, or equipment suppliers, at the regional as well as the national level. This opens the door to many forms of competition and cooperation, from joint-venture partnerships to licensing and distribution agreements, to strategic but often shifting alliances, to outright trade war.

Analysing the dynamics of the telecommunications and media sectors from these two perspectives provides a basis for understanding many of the strategic issues involved. One of the strategic issues, for example, is the often cited need for foreign companies to find an Asian partner to do business in local markets. One reason for this is that local capital may enjoy an access to information not immediately available to a foreign company, an access that comes from business, family or political networks. This reflects a general lack of developed information markets in most Asian economies, and there may be strong local interests wanting to keep it that way. But the thrust of telecommunications is to make information accessible, and this is especially so when network expansion is accompanied by easily acquired end-user terminal equipment, such as telephones, facsimile machines, computers, satellite dishes, radio antenna and so forth. These devices make information accessible, and accessibility leads to demand, from business and social users, and demand stimulates supply. It may be forecast that access to telecommunications will undermine the privileged access to information currently enjoyed by local capital and local family, business and political networks. Absolute asymmetry of information will be replaced by a relative asymmetry, and the multinational corporations in particular will make sure it is not they who are information poor.

Future Research

The chapters of this book have demonstrated that telecommunications development in Southeast and East Asia is not a linear process. New technologies allow countries to leapfrog from non-existent or antiquated analogue systems directly to high technology. Despite the overall low national teledensities across Asia, including regions entirely unserved by a basic telephone, the fast-track of development in the regions most closely integrated with world market encourages a demand for the liberalization of regulations governing equipment, networks and services. This diversity of experience poses questions for future research.

For the majority of people in Asia the issue of universal service is the most pressing, and the most thorny problem. Is it an achievable goal, and can it be achieved only through massive state subsidy? In developing Asian economies the most intractable part of the universal service obligation is rural service. Does the development process change significantly the economics of rural telecommunications? An example of how development may affect rural service provision is the issue of labour migration. Migrant labour is not new to Asia, but the transformation overtaking Asian economies now means that hundreds of thousands, indeed millions, of migrant rural workers are moving to towns and cities, at home and abroad, to become wage-earners with business and social reasons to pay for telephone calls back to their home villages. This creates new traffic and revenue streams with the potential to sustain rural networks. At the same time new technologies, such as radio and satellite communications which digitally assign limited circuits to multiple users,[1] create opportunities for market entry by small entrepreneurs in these local markets.

Development is therefore creating new possibilities on the demand and supply sides of the industry, but sensible government regulation is necessary to realize this potential. In the case of rural telecommunications, for example, new local entrants probably require guaranteed revenue-sharing arrangements with the trunk networks which provide the all-important communications between remote rural networks and the cities and international gateways. Issues such as these, along with their economic and social implications, need detailed investigation across Asia.

New entry, and the conditions governing it, is one issue that requires investigation; the by-pass of public networks is another. What policy measures are appropriate in developing Asian economies to strike a balance between, on the one hand, the objective of overall public network expansion, and on the other, meeting the demands of the large corporate users for lower IDD and long-distance tariffs, self-provisioning and by-pass of the public networks, sharing of leased circuits and the resale of spare capacity on them, and so on? How far and how fast should Asian governments follow the lead of the USA, Britain and other developed countries, in allowing telecommunications carriers, cable TV narrowcasters and broadcasters, and online information service providers to compete in each other's markets? Broad statements of principle are insufficient. Again, detailed studies are required to identify the costs and benefits of alternative courses of action in particular jurisdictions.

Further enquiry into regulation as an issue concerns not just the content of regulation, but also its organization. As telecommunications markets are liberalized there is a growing need for effective and efficient regulation, yet throughout most of Asia few resources are currently made available for regulation, and especially not for the idea of a regulator independent from day-to-day supervision by a government minister. Much more research is required to identify, in the context of Asia's different state regimes, the relationship between the aims of public policy, the objectives of the regulator and the methods of regulation available.

A traditional method of regulation used to protect monopoly operators was to restrict the types of equipment that could be connected to the public switched telephone network. Equipment had to conform to certain standards; by controlling the testing laboratories the monopoly carrier — usually a department of the communications ministry — could also erect very effective non-tariff barriers to trade. Standards issues have always been politically highly charged, and the emergence of Asian equipment and components suppliers are adding to the stakes to be played for. The proliferation of standards remains a major problem for developing economies, which often have to buy what they can from whom they can, rather than what they need from whom they choose. In the coming era boundaries between wireline and wireless networks will dissolve, and this raises many new standards issues concerning emerging regional services such as roaming and international satellite mobile communications. As fast as international agreements on standards can be agreed, another set emerge. In the field of private data networks, for example, the proliferation of standards is giving rise to a whole new business of systems integrators. What model, if any, of integration will emerge across Asia? A political economy of standards in the region remains to be written.

In general, at the policy level, further research needs to be done into the drivers of policy-making in the context of nation-states, and how the influences on policy-making may be changing. The influences arise from domestic and international sources, and these require close examination in the Asian context, especially in light of the growth of intra-regional trade and investment, and of the inter-penetration of capital and migration of people between Southeast and East Asia and other regions.

Undoubtedly one influence is finance. Methods of financing telecommunications network expansion have already received considerable attention from the corporate, securities and banking sector.[2] What has received less attention is the role of local Asian capital markets in mobilizing funds for telecommunications, and the phenomenon of local entrepreneurial businesses entering the telecommunications markets. These are often family-owned businesses with no previous telecommunications experience, but they know how to take advantage of rising stock markets. The telecommunications business requires continuous capital commitment for network expansion and upgrade, and this may not be compatible with market opportunities which are limited nationally to metropolitan areas. On the other hand, region-wide investment opportunities do exist for these companies, and this could have a significant impact on the growth of intra-regional trade and traffic. Expanding geographically is one way to widen the market. Expanding into related information and multi-media markets is another. Much more research is needed to understand the dynamics of these newly emerging local Asian telecommunications and media companies.

Local Asian companies are entering the market just at a time when the convergence of broadcast, computer and telecommunications technologies is prising open the door to multi-media products and services. This makes their market entry strategies particularly interesting to study, but there are problems that need addressing. One is the diversity of Asian cultures, languages, religions and ideologies which subverts attempts to forecast the way in which multi-media services may diffuse. A second is the lack of an infrastructure to provide multi-media transmission or reception. Given the still very low penetration levels of computers and basic telephones, this will be a problem for services which require databases and interactive online modes of usage. By contrast, modes of usage which can make do with terrestrial or satellite radio communications, compact disks, cassettes, television sets or video-recorders, will have fewer problems. A third problem is the general lack of resources available in developing economies for the sectors which in the developed countries have led the way in multi-media applications, such as education, health care and community services. The emerging multi-media industry, in Asia as elsewhere, promises the moon, but what is the *virtual* reality? The most fruitful research on the multi-media industry is likely to be research which disaggregates, which looks at the economic and social implications of particular service applications for their users in the context of the country concerned.

Prior to these immediate policy and market issues, a perspective on Asian development is necessary, which has been an aim of this book. There remains a pressing need to identify and specify more clearly the relationship between the modern telecommunications sector and economic and social development. In particular, research is needed on the pace and patterns of diffusion of telecommunications technologies through Asian economies, and their usage by different economic and social sectors. This information would assist governments, service providers and user groups to prioritize the allocation of resources within the telecommunications industry, for example between types of services and locations, and between telecommunications and other competing sectors such as power-generation and transportation.

Most academic studies have tended to focus on developed countries[3] where the structural and policy issues are very different from those confronting developing countries; for example, where universal service has been more or less achieved and the marginal costs of providing service are close to zero. But the ITU has been active in commissioning studies on telecommunications and development;[4] of particular interest is the volume *Information Telecommunications and Development* (ITU, 1986) which focuses on the importance of identifying the linkages between telecommunications and economic and social usage through the input and output of telecommunications-related services by sector.[5] It is worth quoting from this document:

> Wellenius (1984) pointed out that 'more of the same' studies documenting the relationship between telecommunications and economic development would not help very much to promote accelerated telecommunications investment in developing countries . . . What is needed really is knowledge about the benefits of telecommunications within a specific context in the social, economic and geographic environment of the country in which the decision is to be made. (p.8)[6]

These areas, of course, do not exhaust the research possibilities, but underlying all of them is the need to study, in greater detail than has been done up to now, the demographics of the Asia region, including the cross-migration of people, capital, trade and traffic. These all take place within a context, and as this short conclusion has emphasized, the context is as important to analyse as the content. Only when that level of detailed research has been undertaken will it be possible to make a really informed assessment of the role, impact and future implications of telecommunications development in Asia.

---

[1] Known as DAMA, or digitally assigned multiple access, now widely used in satellite communications.

[2] In recent years the World Bank, and its affiliate the International Finance Corporation, or IFC, have become more involved in lending to state and private telecommunications ventures, with the aim of encouraging sectoral reform, including corporatization and privatization, regulatory reform and tariff reform. See Wellenius, Stern, Nulty and Stern (1989), Ambrose, Hennemeyer and Chapon (1990), Smith and Staple (1994), Wellenius and Stern (1994). See also for the World Bank, Hanna (1991) on information technology and development, and Takano (1992) on the lessons for developing countries of the privatization of NTT of Japan. The Asian Development Bank and the Asia-Pacific Telecommunity have also promoted the issues. See ADB/APT (1991).

[3] See Snow (1986) for a literature review of the literature to the mid-1980s.

[4] The pioneering work was done by Jipp (1963). Later studies up to the early 1980s are summarized for the World Bank by Saunders, Warford and Wellenius (1983, 1994). Studies commissioned by the ITU (International Telecommunications Union) include Pierce and Jequier (1983) and ITU (1986; 1988a; 1988b). The specialist journal *Telecommunications Policy* is a useful source for other studies, including the paper by Di and Liu in Mueller (1994) which reports the internal rate of return on telecommunications investment in China over a ten year period has been estimated at 45 per cent. But there are many hidden subsidies to telecommunications in China, so caution is needed in making judgments.

5   This is a far more useful approach than the conventional adoption of regression equations or production functions because it focuses upon the value-added process which changes over time, and therefore should be capable of identifying and specifying more clearly the changing relationship between telecommunications and real economic growth.
6   On page 9 the aim of the book is defines as follows: 'if . . . telecommunications must develop at a rate that is a mutliple of the rate of general economic expansion, then we have to determine precisely what governs this multiple. The main purpose of this work is to specify this multiple.' This aim remains relevant in the emerging information economies of Asia.

General Agreement on Trade in Services 1994 Annex on Telecommunications

1. *Objectives*

    Recognizing the specificities of the telecommunications services sector and, in particular, its dual role as a distinct sector of economic activity and as the underlying transport means for other economic activities, the Members have agreed to the following Annex with the objective of elaborating upon the provisions of the Agreement with respect to measures affecting access to and use of public telecommunications transport networks and services. Accordingly, this Annex provides notes and supplementary provisions to the Agreement.

2. *Scope*
    (a) This Annex shall apply to all measures of a Member that affect access to and use of public telecommunications transport networks and services.[1]
    (b) This Annex shall not apply to measures affecting the cable or broadcast distribution of radio or television programming.
    (c) Nothing in this Annex shall be construed:
       (i) to require a Member to authorize a service supplier of any other Member to establish, construct, acquire, lease, operate, or supply telecommunications transport networks or services, other than as provided for in its schedule; or
       (ii) to require a member (or to require a Member to oblige service suppliers under its jurisdiction) to establish, construct, acquire, lease, operate or supply telecommunications transport networks or services not offered to the public generally.

3. *Definitions*
   For the purposes of this Annex:
   (a) *Telecommunications* means the transmission and reception of signals by any electromagnetic means.
   (b) *Public telecommunications transport service* means any telecommunications transport service required, explicitly or in effect, by a Member to be offered to the public generally. Such services may include, *inter alia*, telegraph, telephone, telex, and data transmission typically involving the real-time transmission of customer-supplied information between two or more points without any end-to-end change in form or content of the customer's information.
   (c) *Public telecommmunications transport network* means the public telecommunications infrastructure which permits telecommunications between and among defined network termination points.
   (d) *Intra-corporate communications* means telecommunications through which a company communicates within the company or with or among its subsidiaries, branches and, subject to a Member's domestic laws and regulations, affiliates. For these purposes, "subsidiaries", "branches" and, where applicable, "affiliates" shall be as defined by each Party Member. "Intra-corporate communications" in this Annex excludes commercial or non-commercial services that are supplied to companies that are not related subsidiaries, branches or affiliates, or that are offered to customers or potential customers.
   (e) Any references to a paragraph of this Annex includes all subdivisions thereof.

4. *Transparency*
   In the application of Article III of the Agreement, each Member shall ensure that relevant information on conditions affecting access to and use of public telecommunications transport networks and services is publicly available, including: tariffs and other terms and conditions of service; specifications of technical interfaces with such networks and services; information on bodies responsible for the preparation and adoption of standards affecting such access and use; conditions applying to attachment of terminal or other equipment; and notifications, registration or licensing requirements, if any.

5. *Access to and use of Public Telecommunications Transport Networks and Services*
   (a) Each Member shall ensure that service suppliers of any other Member is accorded access to and use of any public telecommunications transport networks and services on reasonable and non-discriminatory terms and conditions, for the supply of a service included in its

schedule. This obligation shall be applied, inter alia, through paragraphs (b) through(f) below. [2]
(b) Each Member shall ensure that services suppliers of any other Member have access to and use of any public telecommunications transport network or service offered within or across the border of that Member, including private leased circuits, and to this end shall ensure, subject to paragraphs 5.5 and 5.6, that such suppliers are permitted:
   (i) to purchase or lease and attach terminal or other equipment which interfaces with the network and which is necessary to supply a supplier's services.
   (ii) to interconnect private leased or owned circuits with public telecommunications transport networks and services or with circuits leased or owned by another service supplier; and
   (iii) to use operating protocols of the service supplier's choice in the supply of any service, other than as necessary to ensure the availability of telecommunications transport networks and services to the public generally.
(c) Each Member shall ensure that service suppliers of any other Member may use public telecommunications transport networks and services for the movement of information within and across borders, including for intra-corporate communications of such service suppliers, and for access to information contained in data bases or otherwise stored in machine-readable form in the territory of any Member. Any new or amended measures of a Member significantly affecting such use shall be notified and shall be subject to consultation, in accordance with relevant provisions of the Agreement.
(d) Notwithstanding the preceding paragraph, a Member may take such measures as are necessary to ensure the security and confidentiality of messages, subject to the requirement that such measures are not applied in a manner which would constitute a means of arbitrary or unjustifiable discrimination or a disguised restriction on trade in services.
(e) Each Member shall ensure that no condition is imposed on access to and use of public telecommunications transport networks and services other than as necessary:
   (i) to safeguard the public service responsibilities of suppliers of public telecommunications transport networks and services, in particular their ability to make their networks or services available to the public generally;
   (ii) to protect the technical integrity of public telecommunications transport networks or services;
   (iii) to ensure that service suppliers of any other Member do not

supply services unless permitted pursuant to commitments in a Member's schedule.

(f) Provided that they satisfy the criteria set out in paragraph (e), conditions for access to and use of public telecommunications transport networks and services may include:
   (i) restrictions on resale or shared use of such services;
   (ii) a requirement to use specified technical interfaces, including interface protocols, for inter-connection with such networks and services;
   (iii) requirements, where necessary, for the inter-operability of such services and to encourage the achievement of the goals set out in paragraph 7(a);
   (iv) type approval of terminal or other requirement which interfaces with the network and technical requirements relating to the attachment of such equipment to such networks;
   (v) restrictions on inter-connection of private leased or owned circuits with such networks or services or with circuits leased or owned by another service supplier; or
   (vi) notification, registration and licensing.

(g) Notwithstanding the preceding paragraphs of this section, a developing Member may, consistent with its level of development, place reasonable conditions on access to and use of public telecommunications transport networks and services necessary to strengthen its domestic telecommunications infrastructure and service capacity and to increase its participation in international trade in telecommunications services. Such conditions shall be specified in the Member's schedule.

6. *Technical Co-operation*

(a) Members recognize that an efficient, advanced telecommunications infrastructure in countries, particularly developing countries, is essential to the expansion of their trade in services. To this end, Members endorse and encourage the participation, to the fullest extent practicable, of developed and developing countries and their suppliers of public telecommunications transport networks and services and other entities in the development programmes of international and regional organizations, including the International Telecommunication Union, the United Nations Development Programme, and the International Bank for Reconstruction and Development.

(b) Members shall encourage and support telecommunications co-operation among developing countries at the international, regional sub-regional levels.

(c) In co-operation with relevant international organizations, Members shall make available, where practicable, to developing countries information with respect to telecommunications services and developments in telecommunications and information technology to assist in strenghthening their domestic telecommunications services sector.

(d) Members shall give special consideration to opportunities for the least developed countries to encourage foreign suppliers of telecommunications services to assist in the transfer of technology, training and other activities that support the development of their telecommunications infrastructure and expansion of their telecommunications services trade.

7. *Relation to International Organizations and Agreements*

(a) Members recognize the importance of international standards for global compatibility and inter-operability of telecommunication networks and services and undertake to promote such standards through the work of relevant international bodies, including the International Telecommunication Union and the International Organization for Standardization.

(b) Members recognize the role played by intergovernmental and non-governmental organizations and agreements in ensuring the efficient operation of domestic and global telecommunications services, in particular for the International Telecommunication Union. Members shall make appropriate arrangements, where relevant, for consultation with such organizations on matters arising from the implementation of this Annex.

---

[1] This paragraph is understood to mean that each Member shall ensure that the obligations of this Annex are applied with respect to suppliers of public telecommunications transport networks and services by whatever measures are necessary.

[2] The term 'non-discriminatory' is understood to refer to most-favoured-nation and national treatment as defined in the Agreement, as well as to reflect sector specific usage of the term to mean 'terms and conditions no less favourable than those accorded to any other user of like public telecommunications transport networks or services under like circumstances'.

Glossary

Access Deficit Shortfall of revenue generated from the provision of access services against the cost of their provision.

Accounting Rate The per minute rate agreed between telephone administrations for the transit or termination of international traffic received. The sending administration pays the receiving administration a certain proportion of this. See Settlement Rate.

ADB Asian Development Bank, based in Manila.

ADC Access Deficit Contribution. A charge levied upon new entrants to compensate the dominant carrier for losses incurred in providing access services.

ADSL Asymmetrical Digital Subscriber Line. A technology for the transmission of compressed digital signals along a twisted-wire pair telephone line. Used in trials for Video-On-Demand. For alternative transmission modes see HFC and FTTH. See also Digital Compression.

AMPS Advanced Mobile Phone System. A North American analogue cellular system typically operating within the 800 MHz frequency.

ANSI American National Standard Institute.

APEC Asia-Pacific Economic Cooperation.

APT Asia-Pacific Telecommunity.

ASEAN Association of Southeast Asian Nations.

ATM Asynchronous Transfer Mode. A broadband switching protocol involving continuous transmit-receive bit sequences, usually of one character length, preceded by a start bit and concluding with an end bit. Suitable for packet switching. Distinguished from synchronous mode.

Audiotex Computer-based stored voice messages accessible over a telephone line.

AUSTEL Australian Telecommunications Authority responsible for the regulation of telecommunications equipment and services.

Bandwidth A range of radio frequencies, measured in Hertz (Hz).

BCC Business Cooperation Contract. A method of financing construction used in Vietnam, where the overseas partner providing equipment, finance and consultancy services is repaid out of the network revenues.

Binary A system of numbers to the base of two; the binary digits being 0 (zero) and 1 (one).

BISDN Broadband Integrated Services Digital Network.

BIT Binary digit.

Bit Rate The speed at which bits are transmitted, usually expressed in bits per second (bps).

BLT Build-Lease-Transfer. A method of financing construction where the builder leases network equipment to the operator. On the expiry of the lease ownership is transferred.

BOT Build-Operate-Transfer. A method of financing construction where the builder owns and operates the facility. On the expiry of the franchise ownership is transferred.

BR (ITU) Radiocommunications Bureau of the ITU, previously the CCIR.

Broadcast Point-to multipoint transmission.

BTO Build-Transfer-Operate. A method of financing construction where the builder transfers ownership but is franchised to operate the facility.

Broadband A channel of radio frequencies wider than a voice circuit 300 Hz to 3,000 Hz.

BTD (ITU) Telecommunication Development Bureau of the ITU.

Busy Hour (BH) The hour of the day during which traffic volumes are greatest. Used for planning network capacity and routing.

Bypass Communications which avoid interconnection with part, or all, of the PSTN.

Byte Eight adjacent bits, or sufficient to represent one alphanumeric character.

C band 4 GHz to 6 GHz radio spectrum used in satellite and microwave transmission.

CAGR Compound Annual Growth Rate.

CATV Community Antenna Television.

CC7 Common Channel Interoffice Signaling. A CCITT digital standard.

CCIR Comite Consultatif de Radio Internationale (Consultative Committee on International Radio) now part of the Radiocommunications Bureau (BR) of the ITU.

CCIS Common Channel Interoffice Signaling (see CC7).

CCITT Comite Consultatif International de Telephonie et de Telegraphie. (International Telegraph and Telephone Consultative Committee) now part of the Telecommunication Standardization Sector (TSS) of the ITU.

CD-ROM Compact Disk-Read Only Memory.

CDMA Code Division Multiple Access. A North American digital cellular standard.

Cellular Radio A local loop network in which each cell is served by its own low-powered transmitter as distinct from mobile cellular telephony which uses either a single high powered transmitter or hand-off between cells.

Central Office (CO) American term for the switching centre which terminates and switches customer circuits. In Europe known as a Central Exchange.

Centrex A PABX facility and service located in the Central Office or Exchange and maintained and operated on behalf of business customers by the local carrier.

Circuit A physical means of directing communication along two or more channels.

Circuit Switching A temporary physical connection between two or more channels, unlike message and packet switching where no direct physical route is set up.

CLI Calling Line Identification. Data generated at the time a telephone call is made identifying the calling party's number; the called party's number; date, time and duration of the call; the routing of the call within and between networks. CLI is an essential element of interconnection agreements between telephone networks.

Closed User Groups (CUG) Communications facilities restricted to the use of members of the CUG.

Coaxial Cable A central wire conductor surrounded by an insulating sheaf and a wire mesh commonly used for radio frequency transmissions 50 MHz to 500 MHz for CATV.

Codec Coder-Decoder used to convert/reconvert analogue to digital signals.

Collection Rate The tariff charged to customers for making international telephone calls. See also Accounting Rate and Settlement Rate.

Common Carrier Provides PSTN and related services.

Communications Satellite Satellite, usually in geostationary orbit 35,800 kilometres (23,000 miles) above the equator to synchronize its rotation with the Earth.

Compression Techniques for conserving bandwidth by reducing the bits required to represent, store or transmit data.

CPE Customer Premises Equipment.

CT-1 Cordless Telephone First generation. Used in the home.

CT-2 Cordless Telephone Second generation. Used in the office and as telepoint outdoors (outgoing calls only).

CT-3 Cordless Telephone Third generation, an Ericsson standard which offers two-way telepoint service.

Data Rate The speed of data transmission in bits per second.

DATEL A European term for data transmission services.

DAX Digital Cross Connect, a switch junction.

DBS Direct Broadcast Satellite.

DCE Data circuit terminating equipment. An access point or network node in a communication system, such as a modem or X.25 connection.

DCS-1800 Digital Cellular System. An ETSI approved standard operating at 1.8 GHz. See PCN.

DECT Digital European Cordless Telephone.

Digital Compression A technqiue for conserving bandwidth by reducing the bits required to represent, store or transmit data.

DTH Direct-To-Home. Satellite TV signals directed to home receiver dishes.

Digital The formatting of information into binary code, '1' or '0' for discrete discontinuous communication, as opposed to analogue formatting which transmits as a continuous waveform.

Digital Data Service (DDS) A digital transmission service supporting speeds up to 1.544M/bps or higher.

Digital Network A network of digital switching and digital transmission.

Digital Switching A switch that uses stored program (computer) control, and replaces frequency-division multiplexing with time-division multiplexing to establish multiple simultaneous incoming and outgoing circuit connections.

Direct Dial In Incoming calls to a PABX are directed to the required station without operator assistance.

Direct Distance Dialing Long-distance calling without operator assistance.

Distributed Data Processing Data processing which is distributed across different work stations and locations and connected by a common communications network.

Distribution Frame The apparatus in a Central Office or Exchange terminating lines from customers.

Downlink The circuit from a satellite to the earth.

DTE Data terminating equipment. End-user terminals and computers that connect to DCE.

Duplex A circuit which allows for two-way communications.

E1-E4 European standard telephone circuit speeds of 2.048 Mbps, 8 Mbps, 34 Mbps, 139 Mbps. Sec T1-T4.

E&M Signaling Pre-digital 'Ear' and 'Mouth' signaling to denote send and receive signals. Being replaced by CCIS signaling for ISDN applications.

E-mail Electronic mail.

Earth Station Ground-based equipment for communication with satellites.

EDI Electronic Data Interchange. An electronic exchange of trading documents which dispenses with the use of paper.

Edifact Electronic Data Interchange for Administration, Commerce & Trade. An EDI standard.

Encryption The scrambling of messages according to a secret code.

Equal Access A regulation requiring a network to provide interconnection to other networks at guaranteed quality and at non-discriminatory prices.

Erlang (Erl) A unit of traffic measurement. One erlang is the intensity at which one telephone circuit would be continuously occupied for one hour, for example, 3600 call seconds per hour generally referred to as 36 CCSs (hundred call seconds).

ETF/POS Electronic Transfer of Funds at Point of Sale. Computer-based shopping system using a bank card for the tranfer of funds from the customer's bank account to the seller's bank account.

ETSI European Telecommunciations Standards Institute.

Exchange See Central Office.

FDMA Frequency-Division Multiple Access. A technique for providing, or switching between, a multiple circuits along a single channel by assigning different frequencies to each.

Fibre Optics Glass fibres used to replace coaxial cables and twisted wire pairs. Digital electronic signals are converted into light by Leds (Light Emitting Diodes) or Lasers (Light Amplification by Stimulated Emission of Radiation) for transmission along the fibres.

FITL Fibre-in-the-Loop. The laying of optical fibre cable in the local telephone loop to provide PSTN customer access.

Flat Rate A fixed charge (which could be zero) for local calls irrespective of time.

FPLMTS Future Public Land Mobile Telephone Service. A future generation of mobile technology under consideration by the ITU/TSB.

Frame Relay High-speed communications technique which sends data in eight bit packet 'frames' used especially between LANS. Unlike X.25 packet switching, frame relay requires predetermined routing, and error detection is replaced by retransmission of entire frames.

Frequency How often a periodic (repetitious) wave form or signal regenerates itself at a given amplitude per second, expressed in hertz (Hz).

Frequency Modulation (FM) A method of modifying a sine wave or 'carrier' signal to make it carry information.

FSS Fixed Satellite Services.

FTTB Fibre-to-the-Building.

FTTC Fibre-to-the-Curb (kerb).

FTTH Fibre-to-the-Home.

Gateway Provides a protocol conversion to allow communications between two networks.

GDP Gross Domestic Product.

GHz Gigahertz (one billion cycles per second).

GNP Gross National Product.

Grade of Service Poisson theory describes the probability that a call will encounter a busy signal during busy hour and provides the basic measure of the quality of a telephone service.

GSM Global System for Mobile (formerly Groupe Speciale Mobile). A European digital cellular standard.

Hand-off Transfer of duplex signaling as a mobile terminal passes to an adjacent cell in a cellular radio network.

Handshaking Exchange of signals between two devices initiating a communications connection.

Hard Wired A wireline communications link that permanently connects two devices.

HDTV High Definition TeleVision.

Hertz (Hz) Unit of electromagnetic frequency equal to one cycle second.

HF High Frequency radio MHz to 30 MHz range, typically used in shortwave radio applications.

HFC Hybrid Fibre/Coaxial. Optical fibre cable in the local loop, and coaxial cable to the home. A higher capacity alternative to ADSL, and a cost-effective alternative to FTTH.

Hub A point at which communications transmissions switch or transit.

I-Series CCITT communications standards and recommendations for ISDN.

IBS International Business Services, satellite-based services at up to 8 Mbps.

IDD International Direct Dialing. Without operator assistance.

IFRB International Frequency Registration Board of the ITU responsible for international radio frequency registration and allocation.

IN Intelligent Network.

Inmarsat International Maritime Satellite. A multilateral organization.

Intelsat International Telecommunications Satellite Consortium. A multilateral organization.

Internet A worldwide computer network. Internet began in 1969 as ARPAnet (Advanced Research Projects Agency of the US Department of Defence) designed for strategic reasons as a distributed network. Through interconnection with universities' networks it has spread to become global, with private operators providing gateway access.

IPLC International Private Leased Circuit. An international telecommunications circuit rented by a PTO to a customer who is usually prohibted to carry third-party traffic. See Simple Resale.

IPO Initial Price Offering.

ISDN Integrated Services Digital Network. CCITT operating standards for the simultaneous digital transmission of various services. Access channels include a basic rate 2B + D (64K + 64K + 16K bps. or 144K bps) and a primary rate that is DS1 (1.544M bps in the US, Japan and Canada and 2.048M bps in Europe).

IT Information Technology.

ITU International Telecommunication Union, a multilateral organization and agency of the United Nations, established to provide standardized communications procedures and practices between different national telecommunications administrations. Now organized into the Telecommunications Development Bureau (BDT), the Standards Bureau (SB) and the Radio Communications Bureau (CRB).

IVANS International Value Added Network Services.

JDC Japanese Digital Cellular telephone. See PDC.

Ku-band 12 GHz to 14 GHz radio spectrum used in satellite and microwave transmission.

L-Band 1 GHz region of radio spectrum used in satellite and microwave transmission.

LAN Local Area Network. Computer network limited to a single site.

LEO Low Earth Orbit Satellite.

LF Low Frequency radio 30 KHz and 300 KHz range.

Local Loop A line connecting a customer's telephone equipment with the local telephone exchange.

MAN Metropolitan Area Network. Computer network providing metropolitan wide access.

MATV Master Antenna Television.

Measured Rate A tariffing system based upon some combination of the number, distance and duration of calls made.

MEO Medium Earth Orbit satellite.

Message Switching A technique that stores messages in queues for each destination, forwarding them when the transmission lines become available.

MF Medium Frequency radio 300 KHz and 3 MHz range.

MHz Megahertz (1 million cycles per second).

Microwave Radio frequencies between 4 – 28 GHz. Used for satellite and line-of-sight terrestial transmission.

MMDS Multi-point Microwave Distribution System.

Modem **Mo**dulator-**dem**odulator for conversion/reconversion of computer digital signals for transmission over telephone lines.

MSC Message Switching Centre.

Multiplex (MUX) Technique to combine two or more transmissions along a single channel.

NAMPS Narrowband Advanced Mobile Phone System.

Narrowband A channel of radio frequencies below the level of a voice circuit 300 Hz to 3,000 Hz, typically using a transmission speeds of between 100 to 200 bps.

NMT Nordic Mobile Telecommunications. A Scandinavian analogue cellular standard typically operating within the 400 MHz or 900 MHz frequencies.

NTSC National Television Systems Committee. A North American TV standard using 525 lines.

OFTA Office of the Telecommunications Authority (Hong Kong).

Online Services Information services accessed from computer databases by dumb or intelligent terminals over telephone lines.

ONP Open Network Provision. A European standard to provide the technical basis of Equal Access.

Optical Fibre See Fibre Optics.

OSI Open Systems Interconnection. A reference model to a universal communications network architecture divided into seven layers: (1) Applications (2) Presentation (3) Services (4) Transport (5) Network (6) Link (7) Physical.

PABX Private Automatic Branch Exchange, typically located on customer premises.

Packet Switching The transmission of data that is switched in grouped binary digits (bits) including call control signals. See Frame Relay and X.25.

PAD Packet Assembler/Disassembler. A device programmed to allow end-user terminals access to a packet switched network.

PAL Phase-Alternative Line. A European TV standard using 625 lines.

PanAmSat Pan-American Satellite.

PBX Private Branch Exchange (Non-automatic).

PC Personal Computer.

PCM Pulse Code Modulation. Used to convert/reconvert analogue to digital signals.

PCN Personal Communications Network. A British term for the European personal communications standard for DCS 1800 typically operating between 1.8 – 2 GHz. See DCS.

PCS Personal Communications Services. A North American personal communications standard based upon several alternative technologies, such as CDMA, typically operating at 900 MHz or between 1.5 – 1.9 GHz.

PDC Personal Digital Cellular. A Japanese wireless access standard typically operating at 800 MHz or 1.5 GHz.

PHP Personal Handy Phone. A Japanese personal communications standard.

PLC Private Leased Circuit. Domestic version of IPLC.

PMR Public Mobile Radio (trunked radio service).

POTS Plain Old Telephone Services.

Private Network Network operated by and for private organizations.

PSDN Public Switched Data Network.

PSTN Public Switched Telephone Network.

PTO Public Telephone Operator.

PTT Postal, Telegraph, and Telephone Organization. Usually a governmental department that acts as the national common carrier.

Public Network Network operated for public access.

Pulse Code Modulation See PCM.

Redundancy Provision of backup capacity in the event of system failure.

Repeater A device to amplify a signal for retransmission.

Resale The retail of capacity that has been purchased from another (wholesale) carrier.

RF Radio Frequency. A frequency usually above 20 KHz.

RHQ Regional HeadQuarters.

RTH Regional Telecommunications Hub.

Satellite Communications See Communications Satellite.

SDH Synchronous Digital Hierarchy. The ETSI equivalent of SONET. An optical transmission standard for broadband communications.

SES Satellite Earth Station.

Settlement Rate The proportion of the accounting rate, usually 50%, paid by the sending telephone administration to the receiving telephone administration. Also known as the Accounting Rate Share. See Accounting Rate.

SHF Super High Frequency. Denotes frequencies from 3 GHz to 30 GHz.

Sideband Frequency band on either side of the carrier frequency, often used by modulation techniques to carry information.

Simple Resale The resale of capacity on a IPLC which is connected to the PTSN at either end.

SMATV Satellite Master Antenna Television.

SMEs Small and medium-size enterprises.

SONET Synchronous Optical Network. An optical transmission standard for broadband communications.

SPC Stored Program Control. Computer programmed switching processor in an electronic (digital) telephone exchange.

Store-and-Forward A service which stores voice or data messages for forward transmission, usually to reduce cost through either bulk or off peak transmissions.

Swift Society for Worldwide Interbank Financial Transactions.

Switched Message Network A service for interconnecting message terminals such as teletypewriters.

Switching Routing transmissions from incoming originating circuits to outgoing destination circuits.

Synchronous Transmission Requires the adjustment of the clock of the receiving terminal to match that of the transmitting terminal. Synchronous transmission is faster than asynchronous transmission because it dispenses with 'start' and 'stop' signal bits. With X.25 DTE connects directly to DCE.

T1 - T4 North American standard telephone circuit speeds ranging from 1.544 Mbps, 6.312 Mbps, 44 Mbps, 273 Mbps (the equivalent of 24, 96, 672 and 4032 voice channels). See E1-E4.

TACS Total Access Communications System. UK analogue cellular system typically operating within the 900 MHz frequency.

Tandem A telephone exchange which interconnects only with local exchanges and not subscribers.

TDMA Time-Division Multiple Access. A technique for providing, or switching between multiple circuits along a single channel by splicing signals into separate time slots.

Teletext Non-interactive broadcast of data, graphics and textual information to dumb terminals for display on subscriber TV screens.

Telex Dial-up stop-start communications over telegraph circuits to teleprinters or other display terminals..

Transmission Speed The speed at which information is transmitted, usually measured in bits per second (bps).

Transponder A receiver/ transmitter repeater on a satellite. Typically a communications satellite will have 12, 24 or 36 transponders.

TSS (ITU) Telecommunication Standardization Sector (TSS) of the ITU, previously the CCITT.

Turnkey System A complete communications system, including hardware and software, assembled and installed by a vendor and sold as a total package.

TVRO Television Receive-Only satellite dish.

Twisted Wire Pair Two insulated wires twisted together to form a customer's telephone line.

Two-Wire Circuit A circuit containing one pair of wires, or radio signals, for one-way, half or full-duplex transmission.

UHF Ultra High Frequency radio 300 MHz to 3 GHz range, typically used for cellular telephones and TV broadcasts.

UMTS Universal Mobile Telephone System. A third generation European cellular concept incorporating intelligent functions.

USO Universal Service Obligation. A requirement on a PTO to provide the basic voice telephone service, as far as possible, upon demand to any subscriber in any part of the territory, either at a uniform tariff, or at one based upon a regulated cost principle.

UPT Universal Personal Telecommunications. A concept of a future universal intelligent personal communications network.

V-series CCITT standards and recommendations for transmission of data over telephone circuits.

Value Added A regulatory designation of services which involve some degree of enhancement by the service provider, are typically differentiated from 'basic' (non-enhanced) services (such as POTS) for purposes of delineating the areas of service in which competitive entry is permitted.

VADS Value Added Data Services. A subset of Value Added Network Services (VANS) or Value Added Services (VAS).

VAS Value Added Services. Services designated as value added and provided over a network which may be operated by a separate entity.

VANS Value Added Network Services. Services designated as value added and provided by the network operator.

VHF Very High Frequency radio 30 MHz to 300 MHz range, typically used for radio and TV broadcasts.

Video Signal Signal frequency in the range 1 MHz to 6 MHz for moving image transmission.

Videoconferencing Multi-party audio-video communications over a telecommunications network.

Videotex Interactive broadcast of data, graphic and textual information to subscriber terminals over the public switched telephone network and adapted television receivers. Also called viewdata.

VIPNET Virtual International Private Network. A regular and committed provision to a customer by a PTO of an international connection over the PSTN as if it were a dedicated physical connection.

Virtual Circuit A transmission path of different circuits that is temporarily dedicated to a customer and appears transparent to a sender who uses a predetermine header for message delivery.

VLF Very Low Frequency radio below 30 MHz.

VOD Video-On-Demand.

Voice Frequency Any frequency within the audio range 300 Hz to 3000 Hz (3 KHz) for voice telephony.

VPN Virtual Private Network. See Virtual Circuit.

VSAT Very Small Aperture Terminal. A small satellite receiver typically used by organizations to communicate between remote sites.

WAN Wide Area Network. Computer network covering more than a single site, potentially global.

Wideband See Broadband.

WLL Wireless Local Loop.

World Numbering Plan CCITT numbering plan that divides the world into nine zones. Each zone is allocated a number that forms the first digit of the country code for every country in that zone. The Zones are as follows: (1) North America, (2) Africa, (3) & (4) Europe, (5) South America, (6) Australia, (7) USSR, (8) North Pacific (Eastern Asia), (9) Far East and Middle East.

V-Series CCITT communication standards and recommendations for exchange and terminal transmission equipment.

X-series CCITT communication standards and recommendations for data and ISDN services.

X.21 Interface between DTE and DCE for synchronous operation on public data networks.

X.25 The interface between DTE and packet-switching DCE.

X.28 The interface between DTE and PADS/DCE for asynchronous operation on public networks.

X.75 The interface between X.25 networks.

X.400 The interface for electronic mail.

References

Aamoth, R. J. 1994. Accounting rate policy impact on telecom carriers in DCs. In *Transnational Data Report* (July/August): 26–34.

ADB/APT. 1991. *The Information Age: Challenges for the Telecommunications Sector in the Asia Pacific Region*. Bangkok: Asian Development Bank/Asia-Pacific Telecommunity.

Ambrose, W.W. and P.R. Hennemeyer, J-P. Chapon. 1990. *Privatizing Telecommunications Systems: Business Opportunities in Developing Countries*. Washington, D.C.: International Finance Corporation.

Ang, Ien. 1991. Global media/ local meaning. *Media Information Australia* (November).

Anuwar, A. 1992. *Malaysia's Industrialization: The Quest for Technology*. New York & Oxford: Oxford University Press.

APEC. 1993. *The State of Telecommunications Infrastructure and Regulatory Environment of APEC Economies*, Volume One. Singapore: Asia Pacific Economic Cooperation.

APEC. 1994a. *The State of Telecommunications Infrastructure and Regulatory Environment of APEC Economies*, Volume Two. Singapore: Asia Pacific Economic Cooperation.

APEC. 1994b. *Telecommunications Working Group Document No: TEL\WG9\PLEN*. Hong Kong: Asia Pacific Economic Cooperation.

APT. 1992. Type approval procedures and processes for telecommunications technical equipment in the Asia Pacific Region. In *Study Group 6: Policies and Regulatory Aspects of Telecommunications Asia Pacific Telecommunity*. Bangkok: Asia Pacific Telecommunity.

APT. 1994. *The APT Yearbook*. Bangkok: Asia-Pacific Telecommunity.

BAH. 1988. *Telecommunications Development in Hong Kong*. London & Hong Kong: Booz, Allen & Hamilton.

Baumol, W. and J. Panzar, R. Willig. 1988. *Contestable Markets and the Theory of Industry Structure*. Cambridge: Cambridge University Press.

Beesley, M. and S. Littlechild. 1989. The regulation of privatized monopolies in the UK. *Rand Journal of Economics* 20.3 : 454–472.

Bruce, R. and J. Cunard, M. Director. 1988. *The Telecom Mosaic: Assembling the New International Structure.* London: Butterworths for the International Institute of Communications.

Bruce, R. and J.P. Cunard. 1994. Restructuring the telecommunications sector in Asia: An overview of approaches and options. In *Implementing Reforms in the Telecommunications Sector: Lessons From Experience.* Eds. Wellenius, B. and P.A. Stern. Washington, D.C.: World Bank.

BTCE. 1991. *Demand Projections for Telecommunications Services and Equipment to Asia by 2010: An Australian Perspective.* Occasional Paper 104 (July). Australia: Bureau of Transport and Communications Economics.

CEPD.1992. *The Six-Year National Development Plan of the Republic of China 1991–1996,* Taipei: Council for Economic Planning and Development for the Executive Yuan.

Chang, W.H. 1989. *Mass Media in China.* Ames: Iowa State University Press.

Chao, T. 1992. Challenges in the development and regulation of New Asia Pacific telecommunications industries and markets.In *Proceedings of the IIR Pan Asian Telecommunications Type Approval and Testing Conference.* Hong Kong, 26 September.

Charmonman, S. 1994. Thailand. In *Telecommunications in the Pacific Basin: an evolutionary approach.* Eds. Noam, E. and S. Komatsuzaki, D.A. Conn. Oxford: Oxford University Press.

Chen, H.T. and C.Y. Kuo. 1985. Telecommunications and economic development in Singapore. In *Telecommunications Policy.* (September): 240–244. London: Butterworth-Heinemann.

Chen, K.Y. and K.W. Li. 1991. Industrial development and industrial policy in Hong Kong. In *Industrial and Trade Development in Hong Kong.* Eds. Chen, K.Y. and M.K. Nyaw, Y.C. Wong. Hong Kong: Centre of Asian Studies, The University of Hong Kong.

Choi, S.K. 1992. Unhappy dragon: telecommunications and Korea's competitive advantage. *Pacific Telecommunications Review* (September): 3–6

Choi, S.K. 1993. The role of information and telecommunications infrastructure in the economic development of Korea. In *Proceedings of the International Conference on Economic Growth and Importance of Information-Telecommunications Infrastructures in Asia.* 18–19 (March):183–204. Tokyo: Institute for Posts and Telecommunications Policy.

Choo, K.Y. and M.K. Kang. 1994. South Korea: structure and changes. In *Telecommunications in the Pacific Basin: An Evolutionary Approach.* Eds. Noam, E. and S. Komatsuzaki, D.A. Conn. Oxford: Oxford University Press.

Chu, G.C. and C. Srivisal, Alfian and B. Supadhiloke. 1985. Rural telephone in Indonesia and Thailand: social and economic benefits. *Telecommunications Policy* v.9.4 (June): 159–69.

Cohen, L. and R. Noll. 1991. *The Technology Pork Barrel*. Washington, D.C.: The Brookings Institute.
Conroy, R. 1992. *Technological Change in China*. Paris: Development Centre of the Organization for Economic Cooperation and Development.
Cowhey, P. and J. Aronson. 1988. *When Countries Talk: International Trade in Telecommunications Services*. Lexington: Ballinger.
Cureton, T. 1992. Providing quality of service to telecommunication customers. In *Proceedings of the IIR Pan-Asian Telecommunications Summit'92*. Westin Stamford Hotel. Singapore.
Day, J.J. 1994. Hong Kong. In *Telecommunications in the Pacific Basin: An Evolutionary Approach*. Eds. Noam, E. and S. Komatsuzaki, D.A. Conn. Oxford: Oxford University Press.
DGT. 1993. *The Sixth National Development Plan: Telecom Modernization Projects*. Taipei: Directorate-General of Telecommunications, Ministry of Transport and Communications.
DOTC. 1993. *Telecommunications Market Profile*. Canberra: Telecommunications Policy Division, Department of Transport and Communications.
Doyle, M. 1992. *The Future of Television: A Global Overview of Programming, Advertising, Technology and Growth*. Lincolnwood: NTC Business Books.
Ducatel, K. and I. Miles. 1992. Internationalization of information technology services and public policy implications. *World Development* 20(12): 1843–1857.
Elbert, B.R. 1990. *International Telecommunication Management*. London: Artech House.
Erbetta, J. 1994. International deregulation — will developing economies suffer? In *Proceedings of the Pacific Telecommunications Conference*. Hawaii: Pacific Telecommunications Council.
Freedom Forum. 1993. *The Unfolding Lotus: East Asia's Changing Media*. New York: The Freedom Forum Media Studies Centre.
Frey, B. 1984. *International Political Economics*. Oxford: Basil Blackwell.
Greenfield, C.C. and E. Lee. 1992. Government information technology policy in Hong Kong. In Government computerization policy on the Pacific Rim: special issue. *Informatization and the Public Sector* v.2.2. Ed. King, J.
Goldstein, H. 1994. The growth of mobile communications in the Asia Pacific Region. In *Proceedings of Financial Times Asia-Pacific Telecommunications Conference*. Hong Kong (March).
Hadden-Cave, P. 1989. Introduction. In *The Business Environment of Hong Kong*. Ed. Lethbridge, D.G. Hong Kong: Oxford University Press.
Hahn, C.K. 1992. Competition and deregulation policies for information communications in Korea. In *Proceedings of the Pacific Telecommunications Conference*. Hawaii: Pacific Telecommunications Council.
Hamilton, A. 1992. The Mediascape of Modern Southeast Asia. *Screen* 33 (1): 81–92.

Hamilton, G.G. 1992. *Overseas Chinese Capitalism*. East Asian Business and Development Working Paper Series. Hong Kong: Centre of Asian Studies, The University of Hong Kong.

Hanna, N.K. 1991. *The Information Technology Revolution and Economic Development*. Washington, D.C.: World Bank.

Hawkins, R.W. 1994. *Regional Technical Infrastructures: The European Position in the Evolving Geo-Politics of Telecommunications*. ENCIP Working Paper. Montpellier: European Network for Communications & Information Perspectives/ Science Policy Research Unit, University of Sussex.

Heng, T.M. and L. Low.1990. The economic impact of the information sector in Singapore. *Economics of Planning* 23: 51–70. Amsterdam: Kluwer Academic.

Howell, J. 1994a. Refashioning state-society relations in China. *The European Journal of Development Research* v.6.1 (June): 197–215

Howell, J. 1994b. Striking a new balance: new social organizations in post-Mao China. *Capital and Class* 54 (Autumn): 89–111.

Huber, P.W. 1987. *The Geodesic Network: 1987 Report on Competition in the Telephone Industry*. Washington, D.C.: US Department of Justice, Antitrust Division.

Huber, P.W. and M.K. Kellogg, J. Thorne. 1993. *The Geodesic Network 11: 1993 Report on Competition in the Telephone Industry*. Washington, D.C: The Geodesic Company.

Hukill, M. 1994. Privatization and regulation of Singapore telecom. *Telecom Journal* v.6.3 (July): 26–30. Bangkok: Asia Pacific Telecommunity.

IFC. 1990. *Privatizing Telecommunications Systems: Business Opportunities in Developing Countries*. Ambrose, W. and P. Hennemeyer, J-P Chapon. Discussion Paper 10: Washington, D.C. International Finance Corporation.

IIC. 1993. *TeleGeography 1993*. Ed. Staple, G. London: International Institute of Communications.

Isaac, G. 1992. *The Indonesian Film and Television Industries — Prospects for Australian Collaboration*. Canberra: Department of Foreign Affairs & Trade.

ITU. 1984. *The Missing Link: Report of the Independent Commission for World-Wide Telecommunications Development*. (The Maitland Commission.) Geneva: International Telecommunications Union.

ITU. 1986. *Information Telecommunications and Development*. Geneva: International Telecommunications Union.

ITU. 1988a. *Contribution of Telecommunications to the Earnings/Savings of Foreign Exchange in Developing Countries*. Geneva: International Telecommunications Union.

ITU. 1988b. *Telecommunications and the National Economy*. Geneva: International Telecommunications Union.

ITU. 1993. *Yearbook of Common Carrier Telecommunication Statistics*. Geneva: International Telecommunications Union.

ITU. 1994. *World Telecommunications Development Report: World Telecommunication Indicators*. Geneva: Telecommunications Development Bureau, International Telecommunications Union.
Jipp, A. 1963. The wealth of nations and telephone density. *Telecommunications Journal*. Geneva: International Telecommunications Union.
Kao, C. 1991. *A Choice Fulfilled: The Business of High Technology*. Hong Kong: The Chinese University Press.
Kaplinsky, R. 1987. *Micro-electronics and Employment Revisited: A Review* Geneva: International Labour Office.
Kim, H. 1992. Financial aspects of Korean telecommunications. *Telematics and Informatics* v.9.1: 13–19.
King, J. ed. 1992. Government computerization policy on the Pacific Rim: special issue. *Informatization and the Public Sector* v.2.2.
Korea Annual 1994. Seoul: Yonhap News Agency.
Kraemer, K.L. and J. Dedrick, S. Jarman. 1994. *Supporting the Free Market: Information Technology Policy in Hong Kong*. The Information Society v.10: 223–246.
Kuntjoro-Jakti, D. and N. Achjar. 1993. *Prospek Ekonomie Indonesia 1992–3*. Jakarta: Kompass.
Kuo, C.Y. 1994. Singapore. In *Telecommunications in the Pacific Basin: An Evolutionary Approach*. Eds. Noam, E. and S. Komatsuzaki, D.A. Conn. Oxford: Oxford University Press.
Lam, V. 1993. Cable telephony — an emerging alternative telecommunications deregulation in Hong Kong. In *Proceedings of the AIC China Telecomms Conference*, 13–15 September, Hong Kong.
Langdale, J.V. 1989. International telecommunications and trade in services. *Telecommunications Policy* (September): 203–221.
Langdale, J.V. 1992. *Regional Administrative Headquarters and Telecommunications Hubs: Australia and the Asia-Pacific Region*. Canberra: Department of Industry, Technology & Commerce.
Lee, S.N. ed. 1995. *Telecommunications and Development in China*. New Jersey: Hampton Press.
Liang, X.J. and Y.N. Zhu. 1994. China. In *Telecommunications in the Pacific Basin: An Evolutionary Approach*. Eds. Noam, E. and S. Komatsuzaki, D.A. Conn. Oxford: Oxford University Press.
Littlechild, S. 1979. *Elements of Telecommunications Economics*. Stevenage: Peter Peregrinus for the Institute of Electrical Engineers.
Lowe, V. 1994. Malaysian and Indonesia: telecommunications restructuring. In *Telecommunications in the Pacific Basin: An Evolutionary Approach*. Eds. Noam, E. and S. Komatsuzaki, D.A. Conn. Oxford: Oxford University Press.
Manapat, R. 1993. *Wrong Number: The PLDT Telephone Monopoly*. Madrid: The Animal Farm Series, Parque del Buen Retiro.

Mueller, M. 1992. *International Telecommunications in Hong Kong: The Case for Liberalization.* Hong Kong: The Chinese University Press/Hong Kong Centre for Economic Research.

Mueller, M. ed. 1994. Telecommunications and the integration of China: special issue. *Telecommunications Policy* v.18.3 (April)

Noam, E. and S. Komatsuzaki, D.A. Conn.eds. 1994. *Telecommunications in the Pacific Basin: An Evolutionary Approach.* New York and Oxford: Oxford University Press.

Park, J.C. 1992. Telecommunications market status in Korea and Dacom's opportunity. In *Proceedings of Commtel'92 International Conference.* Hong Kong: HongKong Telecom Association/HongKong Telecoms Users Group.

Petrazzini, B. 1993. The politics of telecommunications reform in developing countries. *Pacific Telecommunications Review* v.14.3 (March): 4–23.

Petrazzini, B. 1994. Hong Kong's telecom market. *Transnational Data and Communications Report* (July/August): 35–7.

Pierce,W. and N. Jequier. 1983. *Telecommunications for Development.* Geneva: International Telecommunications Union.

Pupphavwsa, W. and D. Stifel. 1993. The Thai economy and communications infrastructure. In *Proceedings of the International Conference on Economic Growth and Importance of Information-Telecommunications Infrastructures in Asia.* 18–19 (March): 207–27.Tokyo: Institute for Posts and Telecommunications Policy.

Ramstetter, E.D. 1993. Prospects for foreign firms in developing economies of the Asian and Pacific region. *Asian Development Review.* v.11.1. Manila: Asian Development Bank.

Redding, G. 1990. *The Spirit of Chinese capitalism.* Berlin: Walter de Gruyter.

Saga, K. 1994. Cost comparison between wireless and wireline systems as local networks. In *Proceedings of the 9th APEC Working Group Meeting on Telecommunications* . Hong Kong (March).

Santosa, S.P. 1993. Technology transfer in the Asia-Pacific Region. In *Proceedings of the Pacific Telecommunications Conference* 1053–1058. Hawaii: Pacific Telecommunications Council.

Saunders, R.J. and J.J. Warford, B. Wellenius. 1983, 1994. *Telecommunications and Economic Development.* Baltimore and London: John Hopkins University Press for the World Bank.

Schiffer, J.R. 1984. Anatomy of Laissez-Faire Government: The Hong Kong growth model reconsidered. In *State Policy, Urbanization and the Development Process: Proceedings of a Symposium on Social and Environmental Development.* Ed. Hills, P. Hong Kong: Centre of Asian Studies, The University of Hong Kong.

Shibusawa, M. and Z. Ahmad, B. Bridges. 1989. *Pacific Asia in the 1990s.* London: Routledge for The Royal Institute of International Affiars.

Shih C.T. and Y. Lin. 1994. *Taiwan: Changes in the Environment for Development*. In *Telecommunications in the Pacific Basin: An Evolutionary Approach*. Eds. Noam, E. and S. Komatsuzaki, D.A. Conn. Oxford: Oxford University Press.

Shirk, S.L. 1993. *The Political Logic of Economic Reform in China*. Berkeley: University of California Press.

Sit, V. and R.D. Cremer, S.L. Wong. 1991. *Entrepreneurs and Enterprises in Macau: A Study of Industrial Development*. Hong Kong: Hong Kong University Press.

Sit, V. and S.L. Wong. 1989. *Small and Medium Industries in an Export-Oriented Economy: The Case of Hong Kong*. Hong Kong: Centre of Asian Studies, The University of Hong Kong.

Smith, P.L. and G. Staple. 1994. *Telecommunications Sector Reform in Asia: Toward a New Pragmatism*. Washington, D.C.: World Bank.

Snow, M. 1986. Telecommunications literature: a critical review of the economic, technological and public policy issues. *Telecommunications Policy* v.10.3 (June).

Son, Y. and S.C. Lee, D.R. Lyi, Y.H. Ju. 1991. Korea's experiences in telecommunications. In *Proceedings of the Pacific Telecommunications Conference*. Hawaii: Pacific Telecommunications Council.

Song, B.N. 1990. *The Rise of the Korean Economy*. Oxford: Oxford University Press.

Staple, G.C. 1992. Winning the global telecoms market: the old service paradigm and the next one. In *TeleGeography 1992*. Ed. Staples, G.C. London: International Institute of Communications: 32–53.

Staple, G.C. and M. Mullins. 1989. Telecom Traffic Statistics — MiTT Matter: Improving Economic Forecasting and Regulatory Policy. *Telecommunications Policy* (June): 105–128.

Steinberg, D.I.1990.*The Future of Burma: Crisis and Choice in Myanmar*. Lanham, New York, London: University Press of America/The Asia Society.

Stuart-Fax, M. 1986. *Laos: Politics, Economics and Society*. London: Frances Pinter.

Subhi, M. 1993. Telekom Malaysia — three years after privatisation. In *Proceedings of AIC China Telecomms Conference*. Hong Kong. (September).

Suehiro, A. 993. Family business reassessed: corporate structure and late-starting industrialization in Thailand. *Developing Economies* 31 (4).

Sung, K.J. 1994. South Korea: Telecommunications policies into the 1990s. In *Telecommunications in the Pacific Basin: An Evolutionary Approach*. Eds. Noam, E. and S. Komatsuzaki, D.A. Conn. Oxford: Oxford University Press.

Takano, Y. 1992. *Nippon Telegraph and Telephone Privatization Study: Experience of Japan and Lessons for Developing Countries*. Washington, D.C.: World Bank.

Tracey, M. 1988. Popular culture and the economics of global television. *InterMedia* 16 (2): 12–17.

Tseng, F.T. and C.K. Mao. 1994. Taiwan. In *Telecommunications in the Pacific Basin: An Evolutionary Approach*. Eds. Noam, E. and S. Komatsuzaki, D.A. Conn. Oxford: Oxford University Press.

Ure, J. 1989. The future of telecommunications in Hong Kong. *Telecommunications Policy* v.13.4 (December): 371-378.

Ure, J. 1992. *The Political Economy of Telecommunications in Hong Kong: Information Technologies and the Management of Change*. Ph.D Thesis. London and Hong Kong: Polytechnic of East London and Centre of Asian Studies, The University of Hong Kong.

Ure, J. 1993. Corporatization and Privatization of Telecommunications in Asean Countries. *Pacific Telecommunications Review* v.15.1 (September): 3–13

Ure, J. 1994a. Telecommunications, with Chinese Characteristics. *Telecommunications Policy* v.18.3 (April): 182–194.

Ure, J.1994b. Financing China's telecoms. In *Proceedings of the AIC 1994 Asia-Pacific Telecommunications China Forum (Beijing)*. Singapore: AIC Conferences.

Ure, J. 1995a. Videotex and information services in Hong Kong. In *Heralding the Information Age: Videotex Development in the Asia-Pacific*. Eds. Kuo, K.C. and C.Y. Ho. Singapore: AMIC.

Ure, J. 1995b. Options and opportunities in China's telecommunications. In *Telecommunications and Development in China*. Ed. Lee,S.N. New Jersey: Hampton Press.

Ure, J. 1995c. Telecommunications: Hong Kong and China after 1997. In *From Colony to SAR: Hong Kong's Challenge Ahead*. Ed. Cheng, J. and S. Lo. Hong Kong: The Chinese University Press

Ure, J. and K.Y. Chen. 1993. Economic growth, industrial structure and telecommunications in Hong Kong. In *Proceedings of the International Conference on Economic Growth and Importance of Information-Telecommunications Infrastructures in Asia* 18–19 (March): 207–27. Tokyo: Institute for Posts and Telecommunications Policy.

Venugopal, P. 1992. Malaysian government computerization policy. In Government computerization policy on the Pacific Rim: special issue. Informatization and the public Sector v.2.2. ed. King, J.

Wade, R. 1990. *Governing the Market. Economic Theory and the Role of Government in East Asian Industrialization*. New Jersey: Princeton University Press.

Wellenius, B. and P. Stern.eds. 1994. *Implementing Reforms in the Telecommunications Sector: Lessons from Experience*. Washington, D.C.: World Bank.

Wellenius, B. 1984. On the role of telecommunications in development. *Telecommunications Policy* v.8.2 (March): 59–66.

Wellenius, B. and P.A. Stern, T.E. Nulty, R.D. Stern. 1989. *Restructuring and*

Managing the Telecommunications Sector: A World Bank Symposium. Washington, D.C.: World Bank

Wong, P.K. 1993. Economic growth and information: telecommunications infrastructures in Singapore. In *Proceedings of the International Conference on Economic Growth and Importance of Information-Telecommunications Infrastructures in Asia* 18–19 (March): 207–27. Tokyo: Institute for Posts and Telecommunications Policy.

Woodrow, B.R. 1991. Tilting towards a trade regime: the ITU and the Uruguay Round Services Trade Negotiations. *Telecommunications Policy* (August): 323–342.

World Bank. 1993. *The East Asian Miracle: Economic Growth and Public Policy.* (World Bank Policy Research Report). Washington, D.C.: Oxford University Press for the World Bank.

World Bank. 1994a. *Telecommunications Sector Reform in Asia: Toward a New Pragmatism.* Smith, P. and G. Staple. Discussion Paper 232. Washington, D.C: World Bank.

World Bank.1989. *Restructuring and Managing the Telecommunications Sector.* Eds Wellenius, B. and P.Stern, T.E. Nulty, R.D.Stern. Washington, D.C: World Bank.

World Bank. 1992a. *Nippon Telegraph and Telephone Privatization Study.* Yoshiro Takano. Discussion Paper 179. Washington, D.C: World Bank.

World Bank. 1992b. *Telecommunications: World Bank Experience and Strategy.* Wellenius, B and Others. Discussion Paper 192. Washington, D.C: World Bank.

World Bank/IFC.1994b. *Telecommunications Sector: Background and Bank Group Issues.* Joint Seminar Presented to Executive Directors. Washington, D.C., USA.

Wu, J.P. 1994. *Economy and Information: Selected Works of Wu Jiapei.* Beijing: Scientific and Technical Documents Publishing House.

Xu, A. 1994. Telecommunications network development in China: growth and challenges and implications for structural reform. In *Proceedings of the Pacific Telecommunications Conference'94.* Hawaii: Pacific Telecommunications Council.

Yu, W.Y. and S. Pahng. 1991. Korea Telecom's globalization scheme in face of restructuring of world telecommunications industry. In *Proceedings of the Pacific Telecommunications Conference.* Hawaii: Pacific Telecommunications Council

Zheng, C. 1994. Computer networking in China. *Telecommunications Policy* v.18.3 (April): 236–42.

Zita, K. 1987. *Modernizing China's Telecommunications: Implications for International Firms.* London and Hong Kong: The Economist Intelligence Unit/ Business International Corporation.

Zita, K. 1993. Price cuts lead to network review. In *Proceedings of the IIR Pan-*

Asian Telecommunications Summit'93. Bangkok: 7–9 December, Regent Hotel.

Zita, K. 1994. China: steps toward political and financial reform. In *Telecommunications in the Pacific Basin: An Evolutionary Approach*. Eds. Noam, E. and S. Komatsuzaki, D.A. Conn. Oxford: Oxford University Press.

Academic and Other Sources of Information

The following are selective lists of English-language journals and newsletters relevant to telecommunications, media and related information technologies in Asia. In the case of newsletters the name of the publisher is given first, followed by some examples of their publications. The editor invites readers to suggest journals and newsletters not included below.

Academic and Official Journals

Asia Journal of Communications
Asian Development Review (ADB)
Communications Magazine (IEEE)
Computer Networks and ISDN Systems (ICCC)
I-Ways (CSIS)
Information and Development
Information Economics and Policy
Informatization and the Public Sector
The Information Society
InterMedia (IIC)
The Journal of Asian Studies
Journal of Communications
Journal of Law and Economics
The Journal of Strategic Information Systems
Media Asia
New Breeze (The New ITU, Japan)
New Era (The Telecommunications Association of Japan)
New Telecom Quarterly
Pacific Telecommunications Review (PTC)
Rand Journal of Economics
Technology and Culture
Telecom Journal (APT)

Telecommunications Journal (ITU)
Telecommunications & Space Journal
Telecommunications Policy
Telematics and Informatics
World Development

Other Journals

Asian Advertising & Marketing (HK)
Asian Business (Singapore)
Asian Communications (UK)
Asia-Pacific Broadcasting (Singapore)
AsiaPacific Space Report (HK)
Asia-Pacific Telecommunications (Singapore)
AsiaWeek (HK)
Business Asia (The Economist Intelligence Unit)
Business China (The Economist Intelligence Unit)
Cellular and Mobile International (US)
China Business Review (US)
China Telecommunications Construction (HK/PRC)
China Trade Report (Far Eastern Economic Review)
Communications International (US)
Communications Networks (US)
CommunicationsWeek International (US)
The Economist (UK)
Far East Business (HK)
Far Eastern Economic Review (HK)
Global Communications (US)
Global Telecoms Business (UK)
The IT Magazine (HK)
I.T. Times (Sin)
International Broadcast Information (HK)
International Cable TV Information (HK)
Media Asia (HK)
Media Information Australia
Multichannel News (US)
Post & Broadcast News (Singapore)
Satellite Communications (US)
Screen Digest (UK)
TelecomAsia (HK)
Telecommunications (US)
Telecommunications Development Asia-Pacific (UK)

Telenews Asia (Australia)
Television Business International (UK)
Variety (US)
Via Satellite (US)
World Broadcast News (US)

Newsletters

PUBLISHER	NEWSLETTER
Baskerville Communications (USA)	Broadcasting & Cables TV International
BIS (Australia)	Mobile & Wireless Communications - Asia Pacific
BRC Consultancy (UK)	World Data Services
BRP Publications (USA)	Telecommunications Report International
Business Monitor International (UK)	China & North Asia Monitor; South East Asia Monitor.
Capital Publications (USA)	Telecom Data Networks
China Telecommunications Construction Publishers (China, HK)	CTC News
Corporate Information Services Ltd (HK)	Mainland Press Digest
East Consulting (Australia)	Asiapac Telecoms
Econmoney (UK)	Telecom Finance
Financial Times (UK)	Asia-Pacific Telecoms Analyst; Mobile Communications; New Media Markets and Satellite TV; etc.
Information Gatekeepers (USA)	China Telecom; Wireless Satellite & Broadcasting; etc.
ITC (USA)	China Telecom Report; Telecom Market Surveys; etc.
Kagan World Media (USA)	Asia Pacific Telecom Investor; Cellular International; etc.
MDIS Publications (UK)	Communications Companies Analysis
Northern Business Information/ Datapro (USA)	Telefacts; Telecom Strategy Letter, Telecom Market Letter
Paul Budde Communciation (Australia)	Telecommunications Strategies Report
Phillips Publishing (USA)	Broadband Networking News; Global Telecom Report; etc.

Probe Research (USA) — Pacific Rim Telecommunications
Pyramid Research (USA) — Telecommunications Development Report
Stuart Corner Information Services (Australia) — Telenews Asia
Telecomeuropa (UK) — GSM Service Monitor
Telecom Publishing Group (US) — Information Networks; Mobile Data Report; Advanced Wireless Communications; etc.
Teleresources (Australia) — Mobiles
TR International (US) — Telecommunications Reports International
TTR (UK) — Telecomms Tariffs Review
Warren Publishing (USA) — Mobile Satellite Reports; Satellite Week; etc.

Source: Mailings; Colin Taylor, IPI Pty Ltd, Surry Hills, NSW, Australia.

Index

ABC Communications (Hong Kong)   73, 82, 98
accounting rate   28, 86, 109, 143
advertisers   159, 160, 162, 175, 187
advertising   111, 153, 154, 159, 160, 168, 177, 178, 182, 185, 187, 189
AFTA (Asian Free Trade Area)   49, 76
Airtouch (USA)   46, 89, 104
AIS (Advanced Information Services; Thailand)   70, 82, 94
Alcatel   18, 55, 73, 200, 202, 204, 205, 206, 207, 212, 215, 219, 220, 221, 225, 233
Ameritech (USA)   84, 90
AMPS   25, 32, 34, 35, 59, 70, 73, 75, 211, 220, 221, 232
APEC (Asia Pacific Economic Cooperation)   76, 115, 210, 232
Apstar   133, 135, 137, 138, 156, 180
Arabsat 1C (Middle East)   137
Argentine   110
Army Telecommunications Company (Vietnam)   74
ASEAN (Association of Southeast Asian Nations)   1, 6, 8, 31, 32, 49, 50, 51, 55, 56, 60, 61, 63, 76, 126, 134, 138, 143, 161, 166, 189
Asia-Pacific Telecommunity (APT)   210, 240
Asian Business News (ABN)   176
Asian Development Bank   46, 71, 223, 240
AsiaSat   58, 74, 79, 132, 133, 135, 137, 138, 144, 154, 157, 158, 170, 175, 187, 188
AT&T (USA)   18, 20, 32, 44, 46, 77, 89, 113, 125, 142, 144, 146, 151, 173, 200, 202, 204, 205, 206, 210, 211, 214, 218, 219, 220, 221, 225, 227, 231
ATM (Asynchronous Transfer Mode)   20, 206, 225, 226, 227, 228
AUSTEL   24
Australia   25, 31, 44, 63, 73, 76, 84, 85, 90, 92, 96, 100, 104, 107, 115, 119, 120, 121, 122, 123, 126, 131, 138, 142, 143, 188, 195, 222, 229
Australian Bureau of Transport and Communications   73, 214, 229

Bandar Seri Begawan   60
Bangkok   2, 67, 68, 69, 70, 94, 95, 105, 153, 156, 157, 163, 175, 180, 223, 224
Bangladesh   8, 57, 85, 86, 92
Batam Island   30, 31, 55
BBC (British Broadcasting Corporation)   76, 157, 158, 159, 163, 167, 170, 187
Beijing   2, 12, 13, 17, 18, 20, 31, 43, 45, 100, 144, 171, 172, 208, 218
Belgium   34
Bell Atlantic (USA)   54, 65, 84, 90, 151
Bell Canada   90
Bell South (USA)   32, 90, 101, 103, 233
Benpres Holdings (Philippines)   63, 103, 167
Bhutan   8

Binariang Sdn Bhd (Malaysia) 59, 96, 100, 135, 188, 189
Britain 25, 27, 84, 152, 153, 237
broadband 20, 24, 30, 37, 39, 69, 114, 150, 172, 173, 196, 206, 225, 226, 231, 235
broadcast 7, 31, 53, 67, 75, 114, 127, 132, 134, 136, 144, 147, 151, 153, 155, 157, 159, 160, 161, 165, 167, 168, 169, 170, 173, 177, 188, 190, 239
broadcasters 134, 136, 154, 158, 175, 182, 186, 236
broadcasting 6, 59, 67, 68, 150, 154, 159, 162, 164, 166, 168, 174, 175, 177, 180, 183, 209, 223, 225
Brooks Telecommunications Corp (USA) 20, 231
Brunei 59, 60, 77, 85, 86, 92, 131
BT (formerly British Telecom) 32, 44, 84, 98, 125, 143, 224
Build-Lease-Transfer (BLT) 107, 108
Build-Maintain-Transfer (BMT) 107
Build-Operate-Transfer (BOT) 21, 29, 65, 108, 112, 117
Build-Own-Operate (BOO) 95, 107
Build-Transfer (BT) 53, 107
Build-Transfer-Manage (BTM) 108
Build-Transfer-Operate (BTO) 53, 66, 67, 68, 69, 70, 94, 107, 108, 110, 219
Burma (Myanmar) 9, 49, 67, 74, 85, 86, 92, 138, 145
Business Cooperation Contract (BCC) 71, 73, 89, 90, 105, 107, 108
business networks 149, 179, 181, 183
by-pass 3, 25, 136, 144, 145, 237

Cable and Wireless (UK) 23, 29, 32, 44, 54, 63, 65, 70, 78, 84, 89, 90, 93, 103, 125, 137, 138, 145, 154
cable television 7, 20, 24, 27, 28, 30, 31, 68, 97, 134, 148, 151, 152, 153, 154, 155, 156, 159, 168, 170, 172, 173, 175, 177, 180, 182, 183, 184, 185, 186, 187, 190, 199, 202, 212, 223, 224, 225, 237
Cambodia 9, 49, 66, 67, 68, 72, 74, 75, 85, 86, 92, 145
Canada 90, 115, 119, 120, 121, 122, 123, 131

Capwire (Philippines) 46, 62, 64, 78, 89, 103
CDMA 35, 212, 220, 232
Celcom (Malaysia) 58, 100
cellular 3, 4, 6, 15, 23, 24, 25, 30, 32, 34, 35, 40, 42, 44, 45, 54, 59, 61, 63, 64, 66, 67, 68, 70, 73, 74, 89, 90, 94, 95, 96, 98, 99, 100, 101, 102, 103, 104, 105, 106, 145, 156, 199, 204, 208, 210, 211, 212, 216, 227, 228, 229, 232, 233
chaebol 35, 179, 184, 186, 191
Champion Technology (Hong Kong) 82, 83, 226
Charoen Pokphand (Thailand) 63, 67, 68, 70, 75, 82, 94, 105, 118, 126, 136, 138, 156, 157, 180, 181, 189, 190, 224
China 1, 2, 9, 10, 11, 12, 13, 14, 15, 16, 17, 18, 19, 20, 22, 23, 24, 28, 29, 31, 32, 38, 41, 42, 44, 48, 50, 52, 57, 59, 67, 68, 73, 74, 75, 77, 80, 85, 86, 90, 92, 96, 100, 101, 106, 107, 110, 113, 115, 116, 118, 119, 120, 121, 122, 123, 126, 129, 132, 134, 135, 137, 138, 141, 143, 144, 145, 147, 148, 149, 152, 156, 160, 161, 162, 163, 164, 169, 170, 171, 172, 173, 174, 175, 179, 180, 182, 184, 187, 189, 190, 193, 195, 196, 202, 203, 205, 206, 208, 209, 214, 215, 217, 218, 219, 220, 222, 224, 225, 226, 227, 228, 230, 231, 232
China Central Television (CCTV) 170, 182
China Telecommunications and Broadcast Satellite Corporation 137
ChinaSat 132, 133, 135
Chunghwa Telecommunications Company (CTC; Taiwan) 40
Cisco Systems (US) 199, 200
CITIC (China International Trust and Investment Corporation) 28, 29, 101, 137, 138, 170
Clark Economic Zone (Philippines) 63, 103
CNN 137, 157, 158, 159, 163, 164, 173, 176, 177, 183, 187
COCOM 217, 227, 233
ColumbiaSat 132, 133, 135
Comlink (Thailand) 67, 70

Communications Authority of Thailand (CAT) 66, 67, 69, 70, 93, 94, 105, 156
Companhia de Telecommunicacoes de Macau (CTM) 29, 90
CompuServe 26, 33
computer 3, 6, 7, 13, 15, 20, 24, 26, 30, 34, 36, 37, 44, 45, 58, 67, 72, 113, 126, 127, 141, 149, 151, 153, 154, 156, 229, 236, 239
computing 6, 150, 201
Comsat (USA) 65, 103
convergence 6, 142, 149, 150, 151, 152, 153, 155, 156, 157, 187, 199, 225, 239
Council for Economic Planning and Development (CEPD; Taiwan) 38, 39, 40, 47
CSL (Hong Kong) 23, 26, 233
CT2 3, 18, 25, 32, 40, 42, 68, 70, 73, 98, 215, 220, 221, 232, 233
Cyprus 188

DACOM (Data Communications Corporation; Korea) 34, 35, 36, 46, 82, 83, 104, 186
data 4, 6, 7, 20, 21, 24, 26, 30, 31, 32, 33, 34, 35, 37, 38, 40, 44, 53, 54, 57, 58, 63, 66, 67, 68, 69, 70, 73, 80, 83, 93, 95, 97, 98, 101, 111, 113, 114, 127, 129, 134, 136, 139, 140, 141, 142, 145, 146, 149, 150, 154, 159, 186, 199, 201, 220, 221, 235, 238
database 13, 14, 20, 26, 27, 33, 34, 36, 38, 65, 112, 127, 140, 142, 239
Demand Assigned Multiple Access (DAMA) 145
Democratic Progressive Party (DPP; Taiwan) 39, 147, 190
Deng Xiaoping 12, 13, 15, 17, 43
Department of Transport and Communications (DOTC; Philippines) 62, 63
Deutsche Telekom 44, 80, 125, 207
Digitel (Philippines) 64, 65, 78, 103
Directorate-General of Posts and Telecommunications (DGPT; Vietnam) 71, 72, 73, 74, 75, 89, 105
Directorate-General of Telecommunications (DGT; China) 13, 14, 135

Directorate-General of Telecommunications (DGT; Taiwan) 38, 39, 40, 47, 196, 205, 225, 226, 227, 228
Director-General of Jabatan Telekom Malaysia (JTM) 56, 58
Directorate-General of Telecommunications (DGT; Indonesia) 52

E-mail 13, 15, 26, 36, 38, 40, 53, 140, 141, 142
Eastern Europe 87, 90, 96
Eastern Telecommunications Philippines Inc (ETPI) 62, 64, 78, 89, 103
Easycall 82, 89, 94
Economic Development Board (EDB; Singapore) 30, 45, 176, 178
Economic Services Branch (ESB; Hong Kong) 24, 24, 154
EDI 15, 26, 27, 30, 33, 36, 38, 47, 58, 141, 142
Egypt 131
enhanced services 6, 112, 114, 140
Enterprise d'Etat des Postes et Telecommunications (Laos) 75
Ericsson (Sweden) 18, 55, 73, 74, 75, 200, 204, 206, 208, 210, 212, 219, 228, 230
ESPN 158, 176, 189
ETACS 32, 59, 221
ETRI (Electronics and Telecommunications Research; South Korea) 34, 35, 37

fax 3, 4, 6, 22, 24, 68, 79, 80, 114, 141, 142, 145, 149, 151, 188, 201, 236
fibre optic, see optical fibre
Fiji 17, 85, 90, 92, 171
finance
  capital 86
  concession 196, 214, 226
  project 105, 215
  soft loan 213, 216, 230, 232
  supplier/vendor 71, 87, 196, 212, 213
Fleet Group (Malaysia) 166, 188
France 63, 119, 120, 121, 122, 123, 131, 142
France Telecom 44, 70, 74, 125, 129, 145

French Polynesia   85, 86, 92
Fujitsu (Japan)   198, 200, 203, 204, 206, 217, 220, 221, 231

GE Information Services (GEIS)   141, 146
General Agreement on Trade-in-Services (GATS)   5, 33, 113, 142
General Agreement on Tariffs and Trade (GATT)   112, 113, 142, *also see* WTO
Germany   119, 120, 121, 122, 123, 211, 212, 227
Globalstar (USA)   35, 46
Globe Telecom (Philippines)   31, 62, 64, 65, 78, 82, 83, 87, 89, 103
GM Hughes (U.S.)   46, 200
growth square   77
growth triangle   55, 60, 77, 78
GSM   19, 25, 30, 31, 32, 43, 54, 55, 59, 70, 73, 75, 77, 96, 98, 100, 112, 209, 211, 212, 220, 221, 232
Guam   85, 86, 92, 131
Guangdong Province (China)   17, 18, 20, 28, 101, 171, 181, 208
Guangxi   17
Guangzhou (China)   2, 12, 13, 20

Hanoi   71, 72, 73, 80, 105
Hawaii   131
Ho Chi Minh City   71, 72, 73, 80, 105, 106
Hollywood   142, 157, 159, 162, 175
Home Box Office (HBO)   158, 176, 177, 178, 182, 187
Hong Kong   1, 2, 4, 5, 6, 9, 11, 12, 14, 18, 19, 22, 23, 24, 25, 26, 27, 28, 29, 31, 32, 34, 38, 43, 44, 77, 82, 84, 85, 89, 90, 92, 93, 95, 97, 100, 104, 107, 114, 115, 116, 117, 118, 119, 120, 121, 122, 123, 126, 129, 131, 132, 134, 139, 141, 143, 144, 146, 147, 148, 149, 152, 153, 154, 155, 156, 157, 158, 159, 160, 161, 162, 163, 172, 174, 175, 176, 177, 178, 179, 180, 181, 187, 189, 195, 202, 210, 211, 220, 224, 226
Hong Kong Cable Communications Ltd (HKCC)   153, 154, 155
Hong Kong Telephone Company (HKTC)   23, 24, 25, 97, 152
Hongkong Telecom (HKT)   23, 24, 26, 28, 32, 36, 44, 82, 83, 89, 90, 93, 97, 98, 107, 125, 155, 176, 233
Hongkong Telecom International (HKTI)   23, 25, 98, 144, 154
hub   17, 31, 68, 129, 134, 139, 171, 178, *also see* regional hub
Hutchison (Hong Kong)   24, 70, 82, 83, 94, 97, 134, 137, 138, 145, 146, 154, 187, 211, 212, 232
Hyundai (Korea)   46

IBM (US)   42, 67, 68, 69, 142, 146, 151, 200
India   8, 57, 68, 74, 82, 85, 86, 90, 92, 106, 107, 108, 120, 122, 138, 145, 158, 160, 161, 187, 195, 199, 202, 220, 228, 233
Indochina   1, 2, 6, 49, 59, 68, 72, 74, 79, 126, 134, 135, 138, 179
Indonesia   2, 6, 9, 30, 31, 46, 50, 51, 52, 53, 54, 55, 57, 60, 62, 68, 76, 77, 78, 79, 86, 89, 92, 98, 106, 107, 110, 115, 116, 118, 120, 121, 122, 126, 127, 131, 132, 135, 136, 141, 143, 145, 147, 148, 158, 160, 161, 162, 163, 165, 168, 174, 180, 187, 195, 202, 204, 207, 210, 220
Infonet (USA)   141, 145, 146
information service   4, 6, 58, 73, 114, 127, 140, 151, 174, 237
Initial Public Offering (IPO)   88, 95, 99, 102, 104, 105, 108
Inmarsat (International Maritime Satellite Corporation)   46, 135, 144
Integrated Services Digital Network (ISDN)   30, 37, 39, 58, 69, 126, 127, 132, 140, 143, 205, 217, 218
Intelsat (International Satellite Organization)   48, 62, 65, 69, 132, 133, 134, 135, 137, 138
International Finance Corporation (IFC)   240
International Telecommunication Union (ITU)   18, 52, 66, 80, 113, 124, 147, 218, 222, 230, 239
international value-added service   142
international value-added services

(IVANS) 6, 15, 25, 139, 140, 141, 142, 145, 146
Internet 26, 72
IranSat 144
Iridium project 35, 46, 48
Israel 160, 161
Italy 63, 119, 120, 121, 123

Jabatan Telekom Brunei (JTB) 60
Jakarta 2, 50, 54
Japan 3, 8, 11, 22, 38, 44, 63, 77, 84, 85, 90, 92, 104, 107, 115, 116, 117, 119, 120, 121, 122, 123, 131, 132, 135, 138, 140, 141, 143, 146, 157, 161, 172, 174, 193, 194, 195, 198, 199, 202, 204, 214, 216, 217, 219, 220, 224, 230, 231
Jasmine (Thailand) 63, 67, 70, 79, 82, 105
Jitong (China) 14, 15, 21, 22, 41, 42, 100, 101
Johor Bahru 2, 55, 56, 78
Joint Operating Scheme (JOS) 51, 55, 99, 107
joint venture 7, 16, 18, 19, 20, 22, 23, 30, 31, 32, 40, 53, 54, 55, 58, 59, 63, 65, 67, 71, 75, 89, 90, 99, 101, 117, 126, 146, 156, 172, 179, 180, 181, 182, 204, 205, 206, 224, 225, 231, 232, 236

KDD (Japan) 32, 44, 125, 144
Kiribati 85, 86, 92
Korea 34, 36, 37, 38, 46, 82, 89, 90, 131, 135, 138, 179, 187
Korea Mobile Telecom (KMT) 35, 38, 46, 47, 82, 83, 104
Korea Telecom (KT) 34, 35, 36, 37, 38, 46, 47, 63, 82, 89, 90, 103, 104, 125, 135, 186, 191, 226
Korean Telecommunications Authority (KTA) 33, 34, 35
Koreasat 37, 133, 135, 191
Kuala Lumpur 2, 56, 59, 175, 188
Kuomintang (KMT; Taiwan) or 'Nationalist Party' 37, 184
Kuwait 161

Laos 9, 49, 66, 67, 72, 74, 75, 76, 85, 86, 89, 92, 145

Liantong (China) 14, 15, 19, 21, 22, 42, 100, 101, 232
Lines Technology (France) 33, 68, 70
Loxley (Thailand) 70, 79, 82, 94, 105, 156, 157

Macau 11, 29, 84, 85, 86, 92, 120, 122, 123
Malaysia 2, 6, 8, 9, 10, 30, 31, 49, 55, 56, 57, 58, 60, 66, 76, 77, 78, 82, 85, 86, 89, 92, 93, 95, 99, 106, 107, 110, 115, 116, 119, 120, 121, 122, 126, 127, 131, 132, 134, 135, 141, 143, 146, 147, 148, 149, 153, 160, 162, 163, 164, 165, 166, 169, 174, 176, 179, 182, 184, 188, 195, 202, 204, 207, 221, 227, 228
Malaysia Telekom 176, 189
Manila 2, 60, 62, 103, 168
Mass Communication Organization of Thailand (MCOT) 68, 187, 223
Matra (France) 33, 200
Matrix (Australia) 70, 89, 94
McCaw (USA) 46, 89, 98, 211
MCI (USA) 44, 125
Measat 59, 78, 100, 133, 135, 144, 145, 188
Mexico 110
Minister of Tourism, Posts and Telecommunications (Indonesia) 50, 52
Ministry of Electronic Industries (MEI; China) 14, 43, 100, 101
Ministry of Energy, Telecommunications and Posts (MEG; Malaysia) 57, 58, 59
Ministry of Posts and Telecommunications (MPT, China) 10, 13, 14, 16, 17, 18, 19, 21, 28, 42, 43, 44, 98, 100, 101, 143, 208, 209, 210, 217, 227, 231, 232, 233
Ministry of Radio, Film and Television (MRFT; China) 20, 21, 170
Ministry of Transport and Communications (MOTC; Thailand) 38, 39, 40, 66, 67, 74, 79
Mobikom (Malaysia) 59, 96, 99
Motorola (USA) 18, 35, 46, 68, 73, 75, 173, 200, 208, 211, 212, 216, 219, 220, 221, 232, 233

multi-media 7, 20, 28, 31, 88, 150, 151, 152, 224, 238, 239

National Telecommunications Commission (NTC; Philippines) 61, 62, 78, 79, 102
NEC (Japan) 18, 77, 198, 200, 203, 204, 205, 208, 216, 217, 219, 220, 221, 231, 232
Nepal 8, 85, 86, 92
Netherlands 119, 120, 121
New T&T (Hong Kong) 24, 97
New World Telephone (Hong Kong) 24, 97
New Zealand 76, 82, 84, 85, 90, 92, 96, 107, 115, 131, 143, 222
News Corporation 158, 181, 190
NHK (Japan) 137, 177, 182, 190
NMT 70, 75, 220, 221
Nokia (Finland) 200, 204
Northern Telecom (Canada) 18, 67, 200, 202, 203, 206, 212, 218, 220, 221, 231, 232
Norway 31
Norwegian Telecom 32
NTT (Japan) 89, 90, 94, 144, 196, 202, 207, 216, 230, 240
Nynex 24, 63, 65, 67, 70, 89, 90, 94, 101, 105, 129, 143, 156, 180, 181, 190, 223, 224, 233

Office of Telecommunications Authority (OFTA; Hong Kong) 24, 25, 98
optical fibre 20, 22, 24, 31, 37, 57, 58, 63, 70, 72, 97, 114, 129, 131, 132, 149, 150, 172, 180, 184, 196
Optus (Australia) 44, 90

Pacific Century (Hong Kong) 41, 94, 134, 144
Pacific Telesis (USA) 90, 94
Pacstar 133, 135
pager 3, 6, 14, 19, 25, 32, 203, 220, 221
paging 4, 15, 24, 31, 34, 35, 47, 53, 54, 63, 64, 66, 67, 70, 73, 75, 83, 89, 90, 94, 98, 101, 106, 136, 156, 227, 232
Pakistan 8, 31, 85, 86, 89, 90, 92, 161, 187

Palapa 53, 54, 62, 79, 99, 132, 133, 134, 135, 137, 138, 145, 157, 158, 163, 188
PanAmSat 133, 135
Papua New Guinea 85, 86, 92
pay TV 148, 163, 166, 184, 186
PBH (Pola Bagi Hasil) 53, 98, 99
PCN 25, 59, 96, 136
Penang 2, 56
Philcom (Philippines) 61, 63, 78, 103
PhilcomSat (Philippines) 62, 64, 103
Philippine Long-Distance Telephone Company (PLDT) 60, 61, 62, 63, 64, 78, 82, 83, 86, 92, 102, 103, 168, 198, 230
Philippines 2, 9, 31, 46, 60, 61, 62, 63, 65, 77, 79, 82, 85, 86, 89, 90, 92, 95, 100, 102, 106, 107, 109, 115, 116, 119, 120, 121, 122, 123, 124, 126, 127, 131, 134, 136, 141, 143, 147, 148, 149, 160, 161, 162, 163, 166, 168, 174, 196, 202, 207, 211, 221, 230
Phnom Penh 67, 75
PHS 216, 217, 233
Piltel (Philippines) 61, 62, 64, 83, 103
President Marcos 60, 65, 78, 167, 168, 196, 230
President Ramos 61, 77, 102
President Suharto 51, 52, 165, 231
Prime Minister Datuk Seri Mahathir Mohamad 59, 76, 77, 162
privatization 6, 30, 34, 40, 66, 84, 94, 95, 104, 240
privatize 35, 51, 57, 69, 104, 105
PT Indosat (Indonesia) 51, 53, 54, 99, 143, 165
PT Pasifik Satelit Nusantara (Indonesia) 135
PT Ratelindo (Indonesia) 54
PT Satelindo (Indonesia) 53, 54, 55, 99, 156, 165
PT Telkom (Indonesia) 30, 51, 53, 54, 55, 75, 98, 99, 165
PT&T (PhilippineTelegraph & Telephone) 46, 63, 64, 65, 89, 103
PTT Telecom (Netherlands) 125
PTT Telecom (Switzerland) 125

Rangoon (Yangon) 74

regional headquarters 139, 140, 175, 176, 177, 178
regional hub 32, 133, 145, 149, 174, 175, 176, 177, 178, 179, *also see* hub
Reinstate 25, 75, 216, 220, 221
Reinstate Television Broadcasts (TVB; Hong Kong) 154, 162, 181
Renong Group (Malaysia) 58, 99, 166
research and development 2, 150, 196, 199, 203, 208, 211, 227, 228, 229, 230
Reuters 26, 141, 176
Rimsat 133, 135, 138
roaming 6, 112, 132, 208, 209
Rupert Murdoch 136, 144, 157, 158, 170, 187, 190
rural 2, 13, 17, 22, 33, 56, 57, 61, 62, 63, 66, 68, 71, 72, 79, 95, 132, 136, 144, 145, 147, 170, 224, 237
Russia 131, 226

Salim Group (Indonesia) 76, 165, 180
Samart (Thailand) 68, 70, 75, 79, 82, 105
Sapura Holdings (Malaysia) 59, 73, 78, 96, 99, 227
satellite 4, 6, 7, 20, 25, 31, 32, 37, 46, 54, 59, 62, 63, 64, 65, 67, 75, 77, 79, 83, 100, 112, 114, 115, 128, 129, 132, 133, 136, 137, 142, 144, 145, 146, 148, 154, 156, 157, 158, 159, 162, 164, 165, 169, 170, 175, 177, 178, 179, 180, 182, 183, 188, 189, 191, 201, 231, 237, 238, 239
satellite broadcast 144, 158, 162, 163, 166, 170, 190, 223
satellite dish 7, 134, 137, 148, 153, 154, 158, 162, 164, 169, 170, 172, 173, 184, 236
satellite television 1, 148, 154, 160, 184, 189, 190
Saudi Arabia 119, 120, 121, 122, 161
SDH (Synchronous Digital Hierarchy) 20, 24, 226, 227
Senior Minister Lee Kwan Yew 29, 45
Seoul 35, 186
Shanghai 2, 12, 13, 17, 20, 45, 100, 101, 110, 131, 171, 172, 180
Shinawatra (Thailand) 31, 66, 67, 68, 70, 73, 75, 79, 82, 83, 89, 94, 103, 105, 126, 134, 135, 138, 145, 156, 180

Siemens (Germany) 200, 203, 204, 205, 207, 219, 220, 221, 225, 230, 232, 233
Singapore 1, 2, 5, 6, 9, 11, 16, 22, 23, 25, 26, 29, 30, 31, 32, 33, 36, 38, 41, 44, 45, 48, 49, 50, 55, 56, 68, 78, 82, 84, 85, 90, 92, 95, 102, 104, 109, 113, 115, 116, 117, 118, 119, 120, 121, 122, 123, 131, 132, 134, 139, 141, 143, 144, 147, 148, 149, 158, 160, 162, 163, 164, 166, 173, 174, 175, 176, 177, 178, 182, 184, 188, 189, 195, 207, 211, 221, 222
Singapore Telecom (ST) 30, 31, 32, 22, 36, 44, 45, 46, 54, 82, 88, 89, 90, 102, 103, 104, 109, 118, 125, 138, 141, 144, 145, 152, 180, 230
Singapore Telecom International (STI) 30, 31, 32, 65, 68, 70, 73, 126, 152
small and medium-sized 23, 27, 29, 44, 117
Solomon Islands 85, 86, 90, 92
SONET *see* SDH
South America 90, 93, 96
South Korea 1, 6, 9, 12, 16, 33, 34, 37, 38, 41, 77, 85, 86, 91, 92, 95, 104, 113, 115, 116, 117, 118, 120, 121, 122, 123, 131, 132, 134, 140, 141, 142, 143, 149, 160, 161, 164, 172, 174, 184, 186, 190, 191, 195, 202, 220, 226
Southwestern Bell (USA) 46, 89, 90, 104
Sri Lanka 8, 31, 85, 86, 92
standards 3, 35, 59, 127, 209, 210, 213, 215, 216, 217, 229, 232, 238
STAR TV (Satellite Television Asian Region; Hong Kong) 137, 144, 154, 156, 158, 159, 160, 161, 162, 165, 170, 172, 175, 182, 183, 187, 188, 190
State Council (China) 13, 14, 16, 19, 42, 101, 169, 171, 175
State Radio Regulation Commission (China) 14, 19, 42, 101, 215, 233
submarine cable 6, 39, 67, 70, 79, 128, 129, 132, 144
Sunkyong (South Korea) 35, 46, 104
Sweden 34
Swedish Telecom 125
Switzerland 119

TACS 25, 75, 216, 220, 221

Taipei 175
Taiwan 1, 6, 9, 11, 14, 16, 32, 36, 37, 38, 39, 40, 41, 47, 48, 77, 85, 86, 92, 113, 115, 116, 117, 118, 119, 120, 121, 122, 123, 129, 131, 134, 137, 138, 139, 141, 142, 143, 144, 147, 148, 149, 158, 160, 161, 162, 163, 174, 175, 178, 179, 182, 183, 184, 187, 190, 195, 196, 202, 205, 217, 221, 225, 226, 227, 228
TDMA 220, 221, 232
Technology Resource Industries Sdn Bhd (TRI, Malaysia) 46, 58, 59, 75, 82, 87, 96, 100, 126
  TRI (Rimsat) 145
technology transfer 7, 12, 16, 22, 51, 96, 126, 173, 202, 203, 207, 208, 218, 228,
Telecom Australia 71, 213, *see also* Telstra
Telecom Denmark 125
Telecom New Zealand 65, 82, 90, 103, 125
TelecomAsia (Thailand) 63, 67, 68, 70, 82, 83, 89, 94, 105, 106, 143, 153, 156, 180, 181, 223, 224
Telecommunications Authority of Singapore (TAS) 30, 104
Teleglobe Inc. of Canada 46, 59
Telekom Malaysia 56, 57, 58, 59, 82, 83, 95, 96, 99, 166, 204
Telephone Organisation of Thailand (TOT) 66, 67, 68, 69, 70, 79, 93, 94, 105
teleport 69, 141
television 75, 112, 132, 134, 147, 148, 149, 150, 151, 152, 153, 155, 156, 159, 162, 163, 164, 165, 166, 168, 169, 170, 173, 174, 178, 180, 182, 186, 188, 190, 191, 223, 224
Television Broadcasts (TVB; Hong Kong) 154, 162, 181
Telstra (Australia) 24, 54, 63, 71, 73, 75, 89, 90, 97, 103, 105, 115, 118, 125, 126, 144, 145, *see also* Telecom Australia
Thai Telephone and Telecommunications (TT&T) 67, 70, 82, 89, 90, 94, 95, 105, 135, 138, 156
ThaiCom 67, 68, 79, 133, 135, 137, 138, 144, 145, 156

Thailand 9, 31, 49, 53, 55, 63, 66, 69, 74, 75, 78, 79, 82, 85, 86, 89, 90, 92, 93, 95, 100, 105, 106, 115, 116, 118, 119, 120, 121, 122, 126, 127, 131, 132, 134, 136, 139, 141, 143, 146, 147, 148, 149, 152, 153, 156, 157, 158, 161, 162, 163, 166, 169, 174, 179, 180, 182, 195, 202, 207, 218, 221, 223, 224
Time (Malaysia) 56, 58, 82, 96, 99
Tokyo 139, 176
Tonga 59, 85, 86, 92
trunk radio 3, 14, 54, 70, 73, 106
type approvals 202, 209, 210, 213

UCOM (United Communications Industry Ltd; Thailand) 66, 68, 70, 82, 94, 105, 183
UMNO (United Malays National Organization) 57, 58, 59, 77, 99, 166
United Arab Emirates 161
United Kingdom 31, 44, 90, 119, 120, 121, 122, 131, 141, 152, 210, 223
United States 3, 11, 25, 31, 34, 35, 38, 46, 49, 71, 76, 90, 115, 119, 120, 121, 122, 123, 124, 131, 141, 142, 143, 152, 158, 159, 177, 183, 199, 204, 210, 211, 212, 216, 218, 231, 237
universal access 102
universal service 5, 25, 58, 95, 236, 239
US Sprint 44, 125, 129, 151
US West 24, 28, 65, 90, 97, 153

value-added services 4, 6, 15, 21, 23, 24, 27, 32, 33, 34, 35, 36, 38, 40, 53, 66, 69, 73, 82, 83, 90, 104, 105, 106, 108, 112, 114, 129, 136, 139, 140, 141, 142
Vanuatu 85, 86, 90, 92
Video-On-Demand (VOD) 7, 28, 30, 142, 148, 153, 155, 169, 223, 225
videotex 15, 33, 36, 37, 68, 70
Vientiane 75
Vietnam 9, 31, 49, 57, 67, 68, 71, 72, 73, 74, 75, 79, 85, 86, 89, 90, 92, 105, 106, 107, 115, 116, 118, 131, 144, 145, 147, 148, 162, 174, 182, 188, 221
VSAT 15, 37, 54, 58, 67, 70, 75, 133, 134, 136, 145

Western Samoa   85, 86, 92
Wharf   24, 97, 153, 155, 158, 172, 175, 187, 224, 225
World Bank   8, 15, 18, 43, 50, 52, 53, 63, 71, 76, 77, 79, 99, 109, 116, 223, 240
World Trade Organization (WTO)   15, 33, 228, *also see* GATT